Argumentation Library

Volume 47

Since 1986 Springer, formerly Kluwer Academic Publishers, publishes the international interdisciplinary journal Argumentation. This journal is a medium for distributing contributions to the study of argumentation from all schools of thought. From a journal that published guest-edited issues devoted to specific themes, Argumentation has developed into a regular journal providing a platform for discussing all theoretical aspects of argumentative discourse. Since 1999 the journal has an accompanying book series consisting of volumes containing substantial contributions to the study of argumentation.

The Argumentation Library aims to be a high quality book series consisting of monographs and edited volumes. It publishes texts offering important theoretical insights in certain major characteristics of argumentative discourse in order to inform the international community of argumentation theorists of recent developments in the field. The insights concerned may pertain to the process of argumentation but also to aspects of argumentative texts resulting from this process. This means that books will be published not only on various types of argumentative procedures, but also on the features of enthymematic argumentation, argumentation structures, argumentation schemes and fallacies.

Contributions to the series can be made by scholars from a broad variety of disciplines, ranging from law to history, from linguistics to theology, and from science to sociology. In particular, contributions are invited from argumentation theorists with a background in informal or formal logic, modern or classical rhetoric, and discourse analysis or speech communication. A prerequisite in all cases is that the contribution involved is original and provides the forum of argumentation theorists with an exemplary specimen of advanced scholarship. The Argumentation Library should enrich the study of argumentation with insights that enhance its quality and constitute a fruitful starting point for further research and application.

All proposals will be carefully taken into consideration by the editors. They are to be submitted in fourfold. If the prospects for including a certain project in the series are realistic, the author(s) will be invited to send at least three representative chapters of their manuscript for review to the editors. In case the manuscript is then judged eligible for publication, the complete manuscript will be reviewed by outside expert referees. Only then a final decision can be taken concerning publication.

This book series is indexed in SCOPUS.

Authors interested in submitting a proposal or completed manuscript can contact either christi.lue@springer.com or the Series Editor.

Barbara Konat

Emotional Appeals in Argumentation

From Rhetoric to Psychology and Artificial Intelligence

Springer

Barbara Konat
Adam Mickiewicz University
Poznań, Poland

ISSN 1566-7650 ISSN 2215-1907 (electronic)
Argumentation Library
ISBN 978-3-032-06331-1 ISBN 978-3-032-06332-8 (eBook)
https://doi.org/10.1007/978-3-032-06332-8

This Springer imprint is published by the registered company Springer Nature Switzerland AG
The registered company address is: Gewerbestrasse 11, 6330 Cham, Switzerland

Preface

My motivation for writing this book is twofold, and both are deeply rooted in my experiences—personal and academic. First, I have observed a growing trend in argumentation theory and informal logic: emotions are no longer dismissed as mere fallacies. Instead, they are increasingly recognized as essential components of real-life persuasion. This shift is visible in both theoretical discussions and empirical research. Second, my own experience of everyday language use diverges starkly from the stylized examples found in argumentation textbooks. These differences are not only linguistic but also functional. We are now in an era where computational linguistics and NLP tools offer new possibilities to collect and analyze natural language at scale. This opens the door to investigating how rhetorical strategies, including emotional appeals, actually function in everyday argumentation.

My deeper, more personal motivation stems from growing up in rural Poland during the 1990s, surrounded by natural beauty but also economic hardship. For me, books became both a form of education and an escape. As a child, I noticed a striking contrast between the dialogues in literature and the actual conversations I heard around me. Literary characters spoke in full sentences, clearly expressing thoughts and emotions, often accompanied by a narrator's inner commentary. In contrast, the spoken language of my community was elliptical, indirect, and socially complex. People rarely communicated in full sentences; instead, they used short phrases, hints, and indirect maneuvers. Emotional appeal was often implied rather than explicitly stated.

This early observation shaped my understanding of language as a tool not only for information exchange but also for action, coordination, and relationship-building. I later learned that these are referred to as the performative, persuasive, and phatic functions of language. These functions are especially prominent in non-academic discourse, and they challenge the idealized view of argumentation centered on rational, logically structured communication.

As I pursued academic studies in philosophy and linguistics, I encountered theoretical models that continued to idealize argumentation. Most textbook examples were prepared, written, and abstract, far removed from the spoken discourse of everyday people. Now, as an academic, I still believe that most argumentation worldwide occurs in informal, spoken settings among people with no formal training in logic or rhetoric. Yet, their persuasive capacities are often remarkably effective. I see it as our scholarly responsibility to understand these practices without prejudice.

This perspective has become especially urgent in recent years. The challenges we face, from hate speech and misinformation to political propaganda, demonstrate that traditional argumentation theory has limited tools to address why people find certain arguments persuasive, even when they are false or dangerous. One reason for this inadequacy, I argue, is our long-standing neglect of the emotional dimension of argumentation.

This book stems from the conviction that a richer and empirically grounded understanding of emotions in argumentation is urgently needed. Through an interdisciplinary engagement with argumentation theory, psychology, and computational linguistics, my aim is to contribute to that understanding.

Poznań, Poland
2025

Barbara Konat

Acknowledgments The writing of this book was supported by a research grant from the Polish National Science Center (Grant no 2020/39/D/HS1/00488), "Com-Pathos: Towards a New Model of Pathos for Computational Rhetoric," and by a sabbatical leave from the Faculty of Psychology and Cognitive Sciences at Adam Mickiewicz University in Poznań. The author declares no competing interests.

Some chapters in this book are based on papers that have either been previously published or are, at the time of writing these acknowledgments, in various stages of publication. Full references are provided where applicable.

The study of metaphors presented as part of **Chap. 2** has been published in:

Juszczyk K, Konat B, Fabiszak M (2025) T*he heart attack of the Polish health service: metaphors, arguments, and emotional appeals in political debates*. Language and Cognition 17:e12, https://doi.org/10.1017/langcog.2024.35

Elements of **Chaps. 3 and 8** were published in:

Konat B, Gajewska E, Rossa W (2024) *Pathos in natural language argumentation: Emotional appeals and reactions*. Argumentation 38:369–403, https://doi.org/10.1007/s10503-024-09631-2

The description of the Emotion Meanings dataset in **Chap. 5** is based on:

Wierzba M, Riegel M, Kocoń J, Miłkowski P, Janz A, Klessa K, Juszczyk K, Konat B, Grimling D, Piasecki M, et al. (2021) *Emotion norms for 6000 Polish word meanings with a direct mapping to the Polish wordnet*. Behavior Research Methods, pp 1–16, https://doi.org/10.3758/s13428-021-01697-0

Chapter 6 describes results of psychological experiments published in:

Konat B, Obrębska M, Gajewska E, Dembska N (2024) *Understanding perceived argument strength: The effects of personal involvement, mood, and pathos appeals*. Submitted Manuscript

Obrębska M, Konat B, Gajewska E, Kleka P, Kiljan K (2024) *Personality traits and argumentation style as factors in perceived argument strength in natural conditions of competitive debating. The role of psychological matching of speaker and listener*. In Press

Chapter 9 contains the results published in:

Gajewska E, Konat B (submitted) *Influencing Hearts: Emotional Appeals in Public Discussions on Social Issues on X (Twitter)*.

Kiljan K, Konat B (In Press) *How to appeal to fear in a fearful context? War proximity and the perception of populists in pre-election debates*.

This research would not be possible without my great team of research assistants in "Computational Pathos" project: Ewelina Gajewska, Nadia Dembska, and Wiktoria Rossa. I would also like to thank the many colleagues who supported me on my journey toward completing this book, through conferences, seminars, and discussions: Katarzyna Adamczyk, Katarzyna Budzyńska, Małgorzata Fabiszak, Martin Hinton, Konrad Juszczyk, Konrad Kiljan, Marcin Koszowy, Dorota Leszczyńska-Jasion, Paweł Łupkowski, Monika Obrębska, Cristian Santibáñez, Menno Reijven, Mariusz Urbański, and Jean Wagemans. I am deeply grateful for the contributions and encouragement of each person mentioned here, whose support brought this work to fruition.

Introduction

"Don't be so emotional; be rational." This is a phrase many of us have either heard or uttered, often assuming that reason and emotion are opposed forces. But is this opposition substantial? Can we be rational without an emotional component? This book proposes a closer examination of the role of emotions in argumentation and persuasion, with the aim of providing a broader theoretical model grounded in interdisciplinary evidence.

This book explores the role of pathos, or emotional appeal, in argumentation by bringing together concepts from argumentation theory, psychology, and computational analysis. It argues for an interdisciplinary approach, presenting a model that incorporates insights from each discipline to analyze emotional appeals in real-world argumentation. The introduction outlines the book's main goals and introduces a framework to examine pathos systematically. This framework is designed to reveal how emotional appeals influence arguments across various discoursive contexts, inviting readers to explore pathos through a comprehensive, empirical lens.

Consider example (1), where the speaker makes an emotional appeal to argue for discontinuing the use of social media:

(1) **Student**: Social media leads to addictions; these are very simple addictions and often not taken as seriously as, for example, alcohol or cigarettes.[1]

The reader may intuitively perceive the presence of an emotional appeal here, and most would probably classify this as evoking negative emotions. The phrase instills fear by presenting something we likely wish to avoid. Some readers might interpret this as eliciting anger, as it points to an obstacle that one might feel compelled to confront. Depending on the reader's predispositions, this emotional appeal could

[1] This example is taken from the corpus of natural language argumentation in a competitive debating experiment presented in detail in Chap. 9. Original in Polish: *Media społecznościowe prowadzą do uzależnień, są to bardzo proste uzależnienia i często nie brane pod uwagę tak poważnie, jak na przykład jakiś alkohol czy papierosy.*

either enhance the argument's perceived persuasiveness or diminish it, prompting skepticism: Why are you appealing to intense emotions instead of providing hard data if you want to persuade me to quit social media?

In this book, I will explore how theoretical and practical tools from argumentation theory, psychology, and AI can help operationalize such intuitions. In example (1), a psychological approach to affective lexicons provides tools for identifying emotionally charged words like "addictions" or "seriously" and for intersubjectively measuring how such words elicit emotions in the general population. Psychology also offers insights into categorical emotions, distinguishing between fear (arising from stimuli we wish to avoid) and anger (triggered by stimuli we want to confront). Through language, these reactions can be evoked symbolically, as specific words substitute for actual stimuli. Part II of this book will elaborate on these psychological tools and findings. The use of computational tools and AI, presented in Part III of this book, can scale up analysis of natural language argumentation, and the use of modern Large Language Models (LLMs) can supplement human analysis effectively.

One of the most infamous examples of emotional appeal can be found in the pre-election debate between Donald Trump and Kamala Harris. During the debate, Donald Trump made an unsubstantiated and false claim in support of tighter immigration control in the United States:

(2) **Donald Trump**: In Springfield, they're eating the dogs, the people that came in, they're eating the cats. They're eating the pets of the people that live there.

This claim, though obviously false and immediately debunked during the broadcast by one of the anchors, was probably effective in eliciting strong emotions—primarily fear, arising from the suggestion of pets being eaten; alienation, from the portrayal of cultural otherness; and intense disgust. Despite widespread criticism of this statement, Donald Trump went on to win the 2024 election, defying predictions that this debate performance would harm his campaign. While multiple factors contribute to a candidate's victory or defeat, it is worth reflecting on how language of this kind—hurtful and offensive—continues to occupy space in public discourse. Emotional appeals in language vary widely in form, ranging from subtle embellishments that enhance an argument to *ad hominem* attacks that border on or cross into hate speech. In Part I of this book we will look at cases of emotional argumentation in pre-election debates to analyze its frequency and use.

This book is not normative in its approach and does not address the ethics of emotional appeals directly. However, Part I will present existing literature on the normativity of emotional appeals, following Walton (1992), to illustrate how certain instances, such as example (1), can be rational and contextually appropriate. In contrast, examples like (2), which spread falsehoods and hatred, demonstrate forms of emotional appeal that should not have a place in public discourse.

The tradition of rhetorical theory, beginning with Aristotle's triad of *logos*, *ethos*, and *pathos*, reminds us that emotional appeals are neither new nor marginal (Aristotle 2004). Modern scholars such as Walton (1992), Macagno (2014), Gilbert (2004), and others have demonstrated that emotions deserve a legitimate place within argumentation theory. Similarly, the French discourse tradition, including scholars like Plantin (2019) and Cigada (2019), has contributed empirical depth to the study of emotional expression in language. Recent developments in computational linguistics further enable the analysis of emotions in large-scale, naturally occurring argumentative discourse (Hidey et al. 2017; Villata et al. 2017).

The structure of the book reflects this integration. Part I provides a historical and theoretical overview of how emotions have been treated in argumentation theory and informal logic. It culminates in my own proposal, the interactional model of pathos (Konat et al. 2024). Part II focuses on the language of emotional appeals and psychological theories of emotion, particularly those relevant to persuasion, such as Ekman's (1999) and Plutchik's (2001) models and the Elaborate Likelihood Model (Petty and Cacioppo 1986). Part III introduces computational linguistics tools for analyzing emotional content and argumentative structure in large datasets (Janier et al. 2014; Visser et al. 2020), illustrating how theory-driven annotation and modeling can be used to operationalize the interactional model of pathos.

In terms of advice for the reader, although the chapters are designed to be self-contained, each with its own bibliography, they are interconnected and build on one another. Both the individual parts and the chapters within them can be read separately, depending on the reader's interests or disciplinary background. However, they are also conceptually linked: all empirical analyses presented in later chapters build on the interactional model of pathos introduced in Part I. Emotional lexicons are introduced in Chap. 5 and later operationalized computationally in Chap. 8. Readers with expertise in specific disciplines may choose to skip certain chapters or focus on those most relevant to their interests. A reader well-versed in argumentation theory might omit Chaps. 1 and 2, while those familiar with psychological models of emotion could bypass Chaps. 4 and 5. Similarly, readers with a computational background may already be familiar with much of the material in Chap. 7. This flexible structure is intended to accommodate diverse readerships while maintaining a coherent theoretical foundation throughout.

This work deliberately avoids making ethical judgments about emotional appeals. While emotions can certainly be used manipulatively, as seen in marketing or propaganda, the book aims instead to document and explain their role in everyday argumentation, especially in spoken and online discourse. The intention is not to prescribe how argumentation *should* work, but to understand how it *does* work in practice. Rather than prescribing how argumentation ought to proceed, this book invites the reader to reconsider what counts as persuasive, by attending to the emotional dynamics that shape real-world reasoning.

References

Aristotle (2004) Rhetoric. Dover Publications

Cigada S (2019) Emotions in argumentative narration: The case of the Charlie Hebdo attack. Informal Logic 39(4):401–431

Ekman P (1999) Basic emotions. In: Dalgleish T, Power M (eds) Handbook of Cognition and Emotion, Wiley Online Library, pp 45–60

Gilbert MA (2004) Emotion, argumentation and informal logic. Informal Logic 24(3)

Hidey C, Musi E, Hwang A, Muresan S, McKeown K (2017) Analyzing the semantic types of claims and premises in an online persuasive forum. In: Proceedings of the 4th Workshop on Argument Mining, Association for Computational Linguistics, Copenhagen, Denmark, pp 11–21, https://doi.org/10.18653/v1/W17-5102

Janier M, Lawrence J, Reed C (2014) OVA+: An argument analysis interface. In: Parsons S, Oren N, Reed C, Cerutti F (eds) Computational Models of Argument, IOS Press, Frontiers in Artificial Intelligence and Applications, vol 266, pp 463–464, https://doi.org/10.3233/978-1-61499-436-7-463

Konat B, Gajewska E, Rossa W (2024) Pathos in natural language argumentation: Emotional appeals and reactions. Argumentation 38:369–403, https://doi.org/10.1007/s10503-024-09631-2

Macagno F, Walton D (2014) Emotive Language in Argumentation. Cambridge University Press, https://doi.org/10.1017/CBO9781139565776

Petty RE, Cacioppo JT (1986) The Elaboration Likelihood Model of Persuasion. Advances in Experimental Social Psychology 19:123–205, https://doi.org/10.1016/S0065-2601(08)60214-2

Plantin C (2019) Tense arguments: questions, exclamations, emotions. Informal Logic 39(4):347–371

Plutchik R (2001) The nature of emotions: Human emotions have deep evolutionary roots, a fact that may explain their complexity and provide tools for clinical practice. American scientist 89(4):344–350

Villata S, Cabrio E, Jraidi I, Benlamine S, Chaouachi M, Frasson C, Gandon F (2017) Emotions and personality traits in argumentation: An empirical evaluation. Argument & Computation 8(1):61–87

Visser J, Konat B, Duthie R, Koszowy M, Budzynska K, Reed C (2020) Argumentation in the 2016 US presidential elections: annotated corpora of television debates and social media reaction. Language Resources and Evaluation 54(1):123–154

Walton D (1992) The Place of Emotion in Argument. The Pennsylvania State University Press, University Park, PA

Contents

Part I Classical Foundations and the Interactional Model of Pathos

1 Classical Rhetoric and Everyday Argumentation 3
1.1 Emotional Appeals as Pathos in Classical Rhetoric 3
1.2 Pathos in Argumentation, Persuasion, and Manipulation 6
1.2.1 Argumentation 6
1.2.2 Persuasion 8
1.2.3 Manipulation 10
References 12

2 The Place of Emotion in Argumentation Theory 15
2.1 Douglas Walton and the Case for Rational Emotional Appeals 15
2.2 Emotive Meaning and Lexical Choice 19
2.3 Other Perspectives 22
2.3.1 Feminist Approaches: Catherine Hundelby 22
2.3.2 Queerness and Emotion: M.A.Gilbert 24
References 28

3 Towards the Model of Interactional Pathos 29
3.1 Cognitive Pragmatics and Relevance Theory 29
3.2 Pragmatics of Emotion 31
3.3 Pragmatics of Argumentation 33
3.3.1 Intersection of Pragmatics and Argumentation 33
3.3.2 Pragma-Dialectics 35
3.3.3 Inference Anchoring Theory 38
3.4 Introducing Model of Interactional Pathos 39
3.4.1 The Model of Interactional Pathos in Argumentation (MIPA) 39
3.4.2 Speech Act Interpretation of MIPA 42

3.5 Model of Interactional Pathos in Argumentation in Pre-election Debates Annotation 45
References 48

Part II Language and Psychology of Pathos

4 The Language of Emotion 53
4.1 Naming Emotions in Argumentation 53
4.2 Emotions in Dialogical Action 55
4.3 Metaphor and Emotions 57
References 60

5 Psychological Models of Emotions 63
5.1 What Is an Emotion? 63
5.2 Categorical Emotions 66
5.3 Psychological Models of Emotion and Argumentation Theory 70
References 71

6 Empirical Applications 73
6.1 Lexicons of Emotion-Eliciting Words 73
6.1.1 Dimensional Models of Emotion in Lexicons 73
6.1.2 Models of Categorical Emotions in Lexicons 76
6.2 Experimental Work in Argumentation Theory 79
6.3 Model of Interactional Pathos in Argumentation for Experiments on Argument Strength 82
6.4 Model of Interactional Pathos in Argumentation for Experimental Competitive Debating 88
References 91

Part III Artificial Intelligence for Pathos

7 AI Approaches to Argumentation 97
7.1 Lexicon-Based Methods in Argumentation Mining 97
7.2 Machine Learning Models for Argument Structure 99
7.3 Large Language Models in Argumentation Mining 100
7.4 Large Language Models in Studies on Emotions in Language 103
References 106

8 Lexicons and Models for Pathos Mining 109
8.1 Using Lexicon-Based Methods for Pathos Mining 109
8.2 Modeling Audience Reactions with Machine Learning 113
8.3 Model of Interactional Pathos in Argumentation in Social Media Discourse 116
References 123

9 Large Language Models for Pathos Mining 125
9.1 Methodology for Pathos Mining with LLM 125

9.2 AI Applications of MIPA in Competitive Debating 129
9.2.1 Quantitative Results in Competitive Debating 129
9.2.2 Case Studies in Competitive Debating......................... 130
9.3 AI Applications of MIPA in Political Discourse 142
9.3.1 Quantitative Results in Political Discourse.................... 142
9.3.2 Case Studies in Political Discourse 145
References 153

Conclusion 155

Index.. 159

Part I
Classical Foundations and the Interactional Model of Pathos

Chapter 1
Classical Rhetoric and Everyday Argumentation

Abstract This chapter introduces the foundational concepts necessary for the study of emotional appeals in natural language argumentation. It begins with the classical notion of pathos in Aristotle's Rhetoric, emphasizing its relevance to contemporary persuasion in both everyday and political contexts. Emotional appeals are examined across traditions including classical rhetoric, informal logic, discourse analysis, and feminist argumentation. The chapter presents pathos as a pragmatically grounded, interactional phenomenon shaped by the speaker to influence the emotional stance of the audience. It distinguishes persuasive argumentation from manipulation and outlines how emotions function within argument structures, not as external accompaniments but as embedded elements. Drawing on Inference Anchoring Theory and contemporary perspectives on persuasion and social influence, the chapter clarifies the role of speaker intent, argument reconstruction, and dialogical context in interpreting emotional appeals. It concludes by situating emotional argumentation as a legitimate and necessary object of empirical study, preparing the conceptual groundwork for the interactional model of pathos and the analyses developed in the following chapters.

Keywords Pathos in rhetoric · Emotional appeals · Persuasive argumentation · Argument reconstruction · Manipulation and persuasion · Inference Anchoring Theory

1.1 Emotional Appeals as Pathos in Classical Rhetoric

Why begin a book that includes psychology and artificial intelligence with Aristotle? While the literature on *Rhetoric* is incredibly rich, I will not attempt to summarize it all. Instead, I focus on two aspects of Aristotle's conception that seem not only strikingly current but above all relevant for this book.

First, I am interested in the very idea of persuasion as something that is constantly present both in everyday life and in political discourse, something which, in both ancient Greece and today, accompanies us if not daily, then at least frequently. This will become evident in the following chapters through the analysis of pre-

B. Konat, *Emotional Appeals in Argumentation*, Argumentation Library 47,
https://doi.org/10.1007/978-3-032-06332-8_1

election debates, student debates, and social media. This allows me to focus the scope of the book to persuasive argumentation, that is, the type of argumentation used to influence someone's beliefs or attitudes. This contrasts with scientific argumentation, which aims to establish the truth of a thesis. The former, as I will show, offers more opportunities to observe emotional appeals. Second, I focus on the concept of pathos as an emotional appeal to the audience's emotions. This idea already leads us toward the insight that emotional appeals are, by their nature, pragmatically grounded and interactional phenomena that presuppose both a speaker and an audience. As we will see later, this is not only foundational to my proposed concept of contemporary pathos, grounded in linguistic theories of pragmatics, but also aligns well with modern psychological understandings of words as linguistic stimuli that elicit emotions.

To begin with, Aristotle (2004), introducing rhetoric as the art of persuasion, leads us towards the realization that persuasion is something every person encounters in their everyday life:

> (...) to a certain extent all men attempt to discuss statements and to maintain them, to defend themselves and to attack others. Ordinary people do this either at random or through practice and from acquired habit. (1354a1-2)[1]

What remains relevant from this citation is, first, the omnipresence of rhetoric. This insight is not confined to ancient oratory, it is also present in everyday life. If the modern reader has opened this chapter after a morning coffee and a quick look at the news, they may already have encountered rhetoric: in a newspaper presenting one stance on current issues as more valuable than another, on social media where influencers propose one product as better than others, or even in a group chat with friends debating where to go camping. This type of everyday rhetoric is what interests me here, the combination of what both "ordinary people" and those with "practice and acquired habit" do to persuade others, and particularly the emotional dimension of that activity. We are bombarded with rhetoric daily. Our decisions, big and small, are influenced by the words of others. This is why I consider this topic so important: combining classical foundations with contemporary technologies can offer a unique perspective on how persuasion works. On the means of persuasion, Aristotle famously writes:

> Of the modes of persuasion furnished by the spoken word there are three kinds. The first kind depends on the personal character of the speaker; the second on putting the audience into a certain frame of mind; the third on the proof, or apparent proof, provided by the words of the speech itself.(1356a4)

These three means of persuasion, later known as logos (argumentation), pathos (emotional appeals), and ethos (speaker credibility), have become foundational to rhetorical theory. In the everyday examples mentioned above, the editorial in a newspaper may lean primarily on logos, providing reasons for a particular stance, but it may also draw upon ethos (credibility of the newspaper) or pathos (invoking

[1] All subsequent references to Aristotle's *Rhetoric* are given in Bekker numbers.

compassion by highlighting the harm done to innocents). Social media influencers rely heavily on ethos, the trust their followers place in them, developed through consistent and engaging content, which they then use to promote products. And among friends discussing camping plans, pathos may dominate, as they appeal to shared memories or emotional bonds to sway the decision. In this book, however, I do not address ethos in detail, as it is the subject of a vast body of separate scholarship and would introduce an additional layer of complexity beyond the scope of the present work.

As my primary interest lies in pathos, let us consider what Aristotle writes:

> Secondly, persuasion may come through the hearers, when the speech stirs their emotions. Our judgements when we are pleased and friendly are not the same as when we are pained and hostile. (1356a5)

This quotation points to the fact that "persuasion may come through the hearers", introducing pathos as a dynamic, interactional process between speaker and audience. Crucially, pathos is not about the speaker's expressed emotions, but about the emotional state of the audience. This distinction is essential, particularly considering Part III of this book, which addresses computational approaches. A common technique in that domain is sentiment analysis: the automatic detection of emotions expressed in text. However, this is not the focus of this study. My interest lies in the emotions of the audience: the persuadee, not the persuader.

An interesting modern perspective on what pathos is comes from Rorty (1984). When analyzing the metaphysical status of pathos, she notes that the term originates in a diffuse, common usage in ancient Greek, generally signifying an experience or event that befalls a person, something passively undergone, usually by accident, as opposed to something actively done. This passive element is central to my approach: pathos is something that is done to the hearer, rather than done by them. According to Rorty, a core characteristic of pathe (plural of pathos) in Aristotle's framework is their exogenous nature, that is, their primary explanation typically lies in an external cause rather than arising from an entity's internal principle of motion. This understanding of pathe as externally caused and passively undergone is crucial in the context of rhetoric. Aristotle notes that a person's thoughts and the reactions that follow from them, can be classified as pathe if they have been externally manipulated by a rhetorician or poet, rather than originating from the person's own activity as a thinking agent. Aristotle assesses the use of such means negatively, claiming that even if the manipulated thoughts are true and the resulting reactions appropriate, such manipulated thought is considered unreasonable. In this book I will refrain from such judgments, simply observing pathos in its natural occurrences and effects.

In light of Aristotle's approach, the emotions of the hearer that influence their judgment can be divided into two types. First are the emotions the audience already possesses, such as the mood or emotional state they bring to the persuasive encounter. In the case of a reader browsing news in the morning, this includes their mood upon waking, the weather, anticipated responsibilities, or interactions with members of their household. These factors lie outside the rhetor's control, but still influence the reception of persuasive messages. In Part II, I will show how such pre-existing emotional states can be captured empirically, for example by asking participants about their mood before engaging with a persuasive task. The second type of pathos, and the reason Aristotle calls it a "means" of persuasion, involves moving the audience through discourse. Aristotle elaborates on how to appeal to specific emotions in later parts of the *Rhetoric*.

When addressing the rhetorical elicitation of specific emotion types, Aristotle focuses on a select few that he considered relevant to the audiences of his time [1378b–1388b]. In one of the modern readings, Kristjánsson (2018) points out that the selection of emotions by Aristotle might be viewed as almost "haphazard" and the reason for this is that he did not aim at a treaty on emotions, but rather a practical guide for orators, listing only those emotions he found useful.

Among the few he selects are anger (and its relation to spite and contempt) and its opposite, calmness. In this book, I choose to use emotion categories grounded in modern psychology rather than Aristotle's original typology, as I believe a broader range of emotional classifications offers deeper insight into the emotional triggers employed by contemporary rhetoric. Nonetheless, there are points of convergence. For example, in Robert Plutchik's classification (discussed in detail in Part II), contempt is understood as a complex emotion (a "dyad") resulting from the combination of two basic emotions: anger and disgust.

To summarize how I adapt Aristotle's ideas, pathos, in this book, refers to what the speaker does with language to affect the emotional stance of the audience for persuasive purposes. I will detail this model in Chap. 3, which provides the conceptual foundation for subsequent empirical work. Following contemporary scholarship and my own research, I will demonstrate how pathos intertwines with logos, i.e. how emotional appeals can be embedded within argument structures either by lexical choices that elicit particular emotions or by modifying argument schemes for that purpose. This leads to a central conclusion: Aristotle's three rhetorical means do not operate in isolation. In practice, a speaker does not necessarily introduce their credibility, then reason through arguments, and finally add emotional appeals. While this sequencing may occur, what interests me here is how emotional appeals are embedded in everyday persuasive argumentation.

It is precisely in this intertwining of and logos, in the ubiquity of persuasion in our lives, and in the concept of pathos as the emotional stance of the hearer shaped by the speaker's words, that the relevance of Aristotle to the work presented in this book becomes clear.

1.2 Pathos in Argumentation, Persuasion, and Manipulation

1.2.1 Argumentation

I based the title of this section on Marek Tokarz's book, *Argumentation, Persuasion, Manipulation* (2006), as I could not think of a better description for the conceptual positioning of pathos, or emotional appeals. In this section, I will briefly introduce the concepts of argumentation, persuasion, and manipulation, as well as my perspective on how pathos fits into each of them. My primary objective here is to present emotional appeals as an inherent element of argumentation and as an area of study that should interest scholars in argumentation theory.

Marek Tokarz, citing Lambert and Ulrich, asserts that the main goals of argumentation are persuasion, hypothesis testing, and explaining phenomena. This perspective incorporates into argumentation a persuasive context, that is, the aim of convincing others. In this book, we will focus almost exclusively on argumentation occurring within a persuasive context, where the objective is to convince someone, such as during televised pre-election debates or experimental settings in which participants aim to win a debate by persuading the audience.

In the context of Aristotle's concept, we can observe that the triad of logos, ethos, and pathos relates to the broader field of persuasion rather than specifically to argumentation (Tokarz 2006, p. 222). Indeed, much research has treated emotions as external elements accompanying arguments rather than being integral to them. I have observed, however, how emotional appeals can be embedded within the linguistic layer of arguments presented in discussions, and this represents the unique perspective proposed in this book. At the same time, this perspective aligns with similar approaches presented later in this book, such as the French school of argumentation analysis focusing on arguing "with" or "about" emotions (Plantin 1999), as well as Walton's views on the place of emotion in argument (Walton 1992). These approaches are complementary and convergent, as despite employing different methodologies, they share the view that emotions can be a part of the argument rather than merely accompanying it.

In this book I adopt a structural concept of argumentation rooted in the Toulminian tradition, defining an argument as a construct that comprises at least one asserted proposition and at least one proposition supporting it. The proposition being supported is referred to as the conclusion, while the supporting proposition is termed the premise. I will return to the question of argument structure in more detail in the sections devoted to Inference Anchoring Theory and Pragma-Dialectics. However two clarifications are necessary here. First, definitions of what constitutes argumentation vary widely. Some focus on criteria related to the quality of what is stated, assessing whether an argument genuinely exists (van Eemeren and Grootendorst 2004). Here, however, we take a different approach, closer to structural reconstruction, without evaluating whether the argument is strong or weak. Instead, we focus on identifying whether the speaker intends for one proposition to support another. This theoretical perspective aligns with Inference Anchoring Theory (IAT) as proposed by (Budzynska and Reed 2011), which reconstructs argumentative and dialogical structures from natural argumentation.

This brings us to a second methodological issue, heavily studied in by IAT. When an analyst reconstructs a proposition and the speaker's intent to provide an argument, the final analysis is necessarily an interpretation. IAT is treating speaker's proposition p not as the proposition itself but as "Speaker John asserts that p." At the level of argument structure reconstruction, this yields the formulation:

"Analyst Bob asserts that Speaker John asserts that *p*" Extending this to argument reconstruction, we observe that the analyst also interprets the speaker's intent to have one proposition support another. Consequently, the reconstructed argument appears as: "Analyst Bob asserts that Speaker John asserts that *p* and that Speaker John justifies *p* with *q*". This format reflects not only the speaker's utterance but also the analyst's interpretative stance toward its argumentative function. And this reconstruction underlies the understanding of argument structure used in this book.

While it would be impractical to fully present examples in this form throughout the book, we should bear in mind that, following IAT, this interpretative framework underlies every reconstruction of argument structure presented here. Neither the interpretation of intent nor the evaluation of emotional appeals should be understood as objective, and the issue of interpretation component in argument reconstruction will not be resolved in this book. It remains a concern for theories of argument reconstruction. Nevertheless, as this work focuses on natural, real-world linguistic arguments, it is essential to remember the distinction between the speaker's intent and the reconstructed intent that emerges in the process of analysis, whether conducted by a human analyst, as explored in Part I, or by a machine, as discussed in Part III. In neither case can objectivity be assumed; interpretation is always at play.

1.2.2 Persuasion

Returning to Aristotle, we can observe that for him, pathos was part of the broader art of rhetoric alongside logos and ethos. In this framework, argumentation might be seen as operating parallel to or in its background. This distinction resembles the modern differentiation between persuasion and argumentation: the former being broader and encompassing the latter alongside other persuasive techniques.

The field of persuasion studies has developed significantly in recent years, primarily within the domain of psychology, with a strong emphasis on experimental methods. O'Keefe (2015) describes persuasion in a way which can be summarized as a successful intentional effort to influence another's mental state through communication in a circumstance in which the persuadee has some measure of freedom. O'Keefe also analyzes the conviction-persuasion distinction (2012) and argues that persuasion entails both influencing attitudes and evoking behavior, combining cognitive and emotional appeals. The traditional alignment of conviction with reason and persuasion with emotion is criticized for conflating communicative means with ends. Instead, O'Keefe proposes disaggregating influence on attitudes from influence on actions, emphasizing that argumentation must often bridge the gap between pre-existing attitudes and desired behavioral change, often through emotional engagement.

Similarly, psychological approaches to persuasion (such as Elaboration Likelihood Model; Petty and Cacioppo 1986), distinguish between pre-existing attitudes and the actual persuasive influence. These approaches are situated within the broader context of social influence and social psychology (and thus also face challenges stemming from the ongoing replicability crisis in social psychology). While this field is dominated by experimental work and often focuses on features describing recipients, such as mood or personal involvement, I will also employ some experimental methods, which are presented in Part II, alongside detailed presentation of ELM. However, my primary emphasis here remains on natural language argumentation, specifically on texts and argument structures.

We can thus see that the concepts of rhetoric and persuasion, while not identical and rooted in distinct traditions and disciplines (philosophy and psychology, respectively) are intimately related. In this book, we aim to untangle the relationship between pathos and argumentation by analyzing both argument structures and linguistic layers, as well as experimentally testing the persuasive power of arguments in practice.

If we define persuasion as an act undertaken to convince someone of a particular belief or action, and an argument as a structure of the form "*p* because *q*" articulated in dialogue, it becomes clear that this book's object of analysis is persuasive argumentation, specifically the role of emotional appeals within it. Importantly, when discussing emotional elements, they are often perceived as external to argumentation, manifesting in the context, appearance, tone, gestures, facial expressions, etc. All these aspects are excluded from this book, which focuses solely on the linguistic and textual expressions of both argumentation and emotional appeals. This narrowing naturally limits the scope of the research and does not aim to comprehensively describe all factors influencing the effectiveness of emotional appeals. However, this simplification allows for a precise analysis of appeals embedded within arguments, rather than those simply accompanying them.

From a pragmatic perspective, an argument becomes persuasive when it involves intention and is realized as a speech act. The relationship between argumentation and pragmatics will be explored in greater detail later in this Part I, where I will present my interactional model of pathos, which is strongly rooted in this approach and builds on Inference Anchoring Theory (IAT). IAT combines Austin and Searle's speech act theory to situate arguments within a dialogical context. As the name suggests, it views arguments as "anchored" in dialogical moves, emphasizing how they are embedded in communicative interactions.

Summarizing what has been said so far, following IAT, I consider "*p* because *q*" as an argument structure. This enables me later to identify, at the linguistic level, additional words eliciting emotions and, at the structural level, specific argument schemes. While the proposed approach does not exhaust all possible considerations regarding emotional appeals in argumentation, I hope this book will demonstrate how unique theoretical, annotative, and computational tools can capture previously unexplored elements of emotional appeals in argumentation.

1.2.3 Manipulation

There is yet a third concept that often comes to mind alongside argumentation and persuasion when discussing pathos: manipulation. As with the previous two, the distinction is not sharp, but it is worth examining situations where eliciting emotions in the audience takes on a manipulative character. While explicitly manipulative tactics are not the focus of this work, it is important to note that scams, frauds, and financial pyramids often employ social engineering techniques based on emotional appeals.

What distinguishes manipulation from persuasion? While the distinction is a vague one, we might assume that the main issue lies in the overt and covert nature of dialogue goals. In persuasion, both sides of the dialogue are aware that persuasion is taking place, whereas in manipulation, the dialogue goal is hidden behind false goals, cases of which will be discussed in later paragraphs.

The reader might notice that this resembles the differences between a sound argument and a fallacious one an argument that, as Walton puts it "looks like a valid argument but is not" (1995). Walton relates this concept to Pragma-Dialectic notions of dialogue goals. Based on this, Walton develops the idea that not all emotional appeals in argumentation are fallacious. They can be reasonable if they align with the overtly stated aims of the dialogue. The use of emotional argument is fallacious only if it leads to a hidden shift in dialogue type and dialogue goals. This analogy between distinguishing fallacious and correct arguments and distinguishing persuasion and manipulation is significant. It all lies in the overtness of dialogue goals. For example, if I use argumentation in a persuasive context, where both you and I know my aim is to persuade you, this is a clear case of persuasion. However, if I pretend that the context is entirely different, for instance, claiming my goal is to help or befriend you, this constitutes manipulation. While emotional appeal can be present in both cases, it raises ethical concerns only in the latter scenario, where its effects can be serious and sometimes deadly.

If we look for classic rhetorical examples of manipulative techniques, *The Art of Being Right by Schopenhauer* (1896) offers a collection of dishonest persuasive methods, including, for example, intimidating an opponent. These techniques, while strategically complex, serve as a useful lens for understanding manipulation's mechanisms more broadly. This consideration leads directly to the topic of manipulation in practical contexts, especially in cases such as scams and cults. Here we can start with perhaps the most famous popular book on influencing people, Cialdini's work (2001). While often associated with the subtitle "psychology of persuasion", it contains techniques that would be called manipulative in the conceptualization proposed here, where part of the success lies in hiding the goal from the persuadee. Although Cialdini's best-selling book focuses on persuasion with only some manipulation elements, it has served generations of manipulators in trade, advertising, and direct sales.

Manipulation is sometimes called the *Dark Persuasion*, and the book with this exact title (Dimsdale 2021) examines the early days of persuasion, its darkest eras under totalitarian systems, and its modern manifestations in online cyberbullying, hate speech, and scams. The book explores both Stalin's use of propaganda and the role of emotional contagion in modern social media influencing users worldwide (Coviello et al. 2014). I share with the author the belief that despite modern tools, the underlying mechanisms remain the same. This belief supports my conviction that we can effectively apply Aristotle's concepts and classical psychological theories to study emotional appeals in modern argumentation.

A startling illustration of manipulation's consequences comes from Steven Hassan's *Combating Cult Mind Control* (2018). A former cult member himself, Hassan provides numerous case studies and a daunting array of techniques used by cults to "brainwash" their members: deprivation of sleep and food, grueling work, and enforced marriage and parenthood within the cult to deter members from leaving. These horrific practices begin with clear cases of manipulation, misleading about the overt and covert goals of the conversation. No one recruits members with the candid claim, "Join our cult, and we will starve you and take your money". Recruiters conceal their actual goals behind the pretense of friendship, romance, or spiritual growth opportunities, often using emotional appeals. Such cases are clearly manipulative: the dialogue's actual goal is hidden from the recipient, and the sole beneficiary is the cult, while the recruited person loses their freedom, money, health, and sometimes their life.

The mechanisms underpinning such manipulation are the same as those in internet scams. In the 2020s, thousands fall victim to scams daily, including romance scams and fraudulent schemes involving impersonation of government authorities or family members. Senior citizens in Western countries are the primary targets (Shang et al. 2022), but even tech-savvy individuals can fall prey because these scams rely on age-old persuasive mechanisms rather than technology. For instance, in romance scams, scammers pretend to seek a romantic relationship, building rapport and trust over time to exploit victims for financial gain. While these cases are tragic and often lead to devastating outcomes, including loss of life, the scammers themselves are frequently victims of human trafficking (Office of the United Nations High Commissioner for Human Rights (OHCHR) 2023). Returning to (Dimsdale 2021), the manipulative techniques used in scams might seem different on the surface from Stalinist propaganda, with online relationships and trafficked human labor versus political actions, but they share the same foundation: concealed dialogue goals, lies, coercion, and emotional appeals.

While a deeper analysis of these cases is beyond the scope of this work, and they each have their dedicated literature, my goal is to demonstrate that while the distinction between manipulation and persuasion is difficult to define in theoretical terms, it becomes evident in specific case analyses. This mirrors Walton's analysis of emotional arguments: while it may not be possible to categorically declare certain arguments fallacious, it is feasible to identify fallacious uses as those causing illicit shifts in dialogue goals (1992). Similarly, this book focuses on analyzing emotional appeals in arguments used in persuasive, not manipulative, contexts. In the analyzed cases of pre-election debates or competitive debates, all parties are aware of the persuasive nature of the situation, enabling observation of their use

Table 1.1 Comparison of argumentation, persuasion, and manipulation

Aspect	Argumentation	Persuasion	Manipulation
Goal	Truth and resolution	Influence and alignment	Exploitation
Logic	Central and explicit	Complementary to emotions	Obfuscated
Ethics	Cooperative	Variable	Unethical
Transparency	High	Moderate	Low

and audience reactions. Analysis of manipulative uses of emotional appeals is left to other studies, though, as we will see in qualitative analyses, some of our cases border on manipulation. I hope that the work presented here will also contribute to a better understanding of these mechanisms. Despite the lack of clear distinctions, a simplified summary of argumentation, persuasion, and manipulation is presented in the Table 1.1.

To conclude, the cases analyzed in the present book are arguments presented in the context of discussions, where an actual or presumed opponent presents (or could present) theses opposing those presented by the speaker (proponent). In the pre-election debates discussed in Parts I and III, the speakers' goal is to persuade the audience while arguing against the theses presented by their opponents. In the experiments described in Part II, the context involves a university attempting to persuade students. In another experiment, students seek to convince an audience of their theses in opposition to counter-theses expressed by an opposing side, within a competitive British parliamentary-style debate organized for this purpose. All this falls into concept of persuasive argumentation, but not manipulation.

The focus on persuasive argumentation also means that other forms of argumentation are excluded from this study, particularly explanatory or epistemic forms, such as those aimed at hypothesis testing or explaining phenomena. The rationale for this exclusion is that emotional appeals are unlikely to play a prominent role in such contexts, especially academic or philosophical ones. However, as noted, this remains an intuition, and one could argue that even academic study reports occasionally contain emotional elements. These elements, if present, would still belong to the persuasive aspects of such texts, despite their primary epistemic aims. Studying such uses of argumentation lies beyond the scope of this book. Nevertheless, I hope that future research will explore emotional appeals in these other domains of argumentation.

References

Aristotle (2004) Rhetoric. Dover Publications

Budzynska K, Reed C (2011) Speech acts of argumentation: Inference anchors and peripheral cues in dialogue. In: Computational models of natural argument: papers from the 2011 AAAI Workshop, AAAI Press, pp 3–10

Cialdini RB (2001) Influence: The Psychology of Persuasion, revised edn. Harper Business, New York

Coviello L, Sohn Y, Kramer ADI, Marlow C, Franceschetti M, Christakis NA, Fowler JH (2014) Detecting emotional contagion in massive social networks. PLOS ONE 9(3):e90315, https://doi.org/10.1371/journal.pone.0090315

Dimsdale JE (2021) Dark Persuasion: A History of Brainwashing from Pavlov to Social Media. Yale University Press, New Haven, CT, twarda oprawa – 28 września 2021

van Eemeren FH, Grootendorst R (2004) A Systematic Theory of Argumentation. The Pragma-Dialectical Approach. Cambridge University Press, Cambridge

Hassan S (2018) Combating Cult Mind Control. Freedom of Mind Press

Kristjánsson K (2018) Virtuous Emotions. Oxford University Press, Oxford

Office of the United Nations High Commissioner for Human Rights (OHCHR) (2023) Hundreds of thousands trafficked to work as online scammers in southeast Asia, says un report. https://www.ohchr.org/en/press-releases/2023/08/hundreds-thousands-trafficked-work-online-scammers-se-asia-says-un-report, accessed: 2025-01-17

O'Keefe DJ (2012) Conviction, persuasion, and argumentation: Untangling the ends and means of influence. Argumentation 26(1):19–32

O'Keefe DJ (2015) Persuasion: Theory and Research, 3rd edn. SAGE Publications, Inc, Thousand Oaks, CA, https://us.sagepub.com/en-us/nam/persuasion/book237644

Petty RE, Cacioppo JT (1986) The Elaboration Likelihood Model of Persuasion. Advances in Experimental Social Psychology 19:123–205, https://doi.org/10.1016/S0065-2601(08)60214-2

Plantin C (1999) Arguing emotions. In: Proceedings of the Fourth Conference of the International Society for the Study of Argumentation, Amsterdam, NL

Rorty AO (1984) Aristotle on the metaphysical status of "pathe". The Review of Metaphysics 37(3):521–546, http://www.jstor.org/stable/20128048

Schopenhauer A (1896) The Art of Being Right. Swan Sonnenschein, London, original work published in 1831

Shang Y, Wu Z, Du X, Jiang Y, Ma B, Chi M (2022) The psychology of the internet fraud victimization of older adults: A systematic review. Frontiers in Psychology 13:912242, https://doi.org/10.3389/fpsyg.2022.912242, published in the section: Psychology of Aging. Part of the Research Topic: Age Differences in Trust and Fraud.

Tokarz M (2006) Argumentacja, perswazja, manipulacja [Argumentation, persuasion, manipulation]. Gdańskie Wydawnictwo Psychologiczne, Gdańsk

Walton D (1992) The Place of Emotion in Argument. The Pennsylvania State University Press, University Park, PA

Walton D (1995) Appeal to pity: A case study of the argumentum ad misericordiam. Argumentation 9(5):769–784

Chapter 2
The Place of Emotion in Argumentation Theory

Abstract This chapter revisits Douglas Walton's foundational work *The Place of Emotion in Argument* and situates it within contemporary developments in argumentation theory, cognitive pragmatics, and empirical discourse analysis. Walton's concept of rational emotional appeals challenges the traditional opposition between reason and emotion, showing that emotional arguments can be normatively acceptable when they align with dialogue goals and rules. The chapter explores how Walton's ideas, grounded in Pragma-Dialectics, have inspired further research into the integration of emotion within natural language argumentation. It examines the emotive potential of lexical choice, drawing on later developments by Walton and Macagno, while also distinguishing between emotional expression and deliberate emotional appeals. The chapter then expands this theoretical foundation through feminist and queer approaches, which critique the gendered assumptions embedded in the rational-emotional dichotomy. The work of Catherine Hundleby and M.A. Gilbert is discussed to highlight how emotion can function as both a communicative mode and a socially situated act. By combining these perspectives, the chapter provides the grounds for an interdisciplinary, empirically informed account of emotional appeals in real-world discourse, laying the conceptual groundwork for the interactional model of pathos developed in subsequent chapters.

Keywords Rational emotional appeals · Douglas Walton · Emotive language · Feminist argumentation theory · Multi-modal argumentation · Pathos in discourse

2.1 Douglas Walton and the Case for Rational Emotional Appeals

The role of emotion in argumentation was famously addressed by Douglas Walton in his book *The Place of Emotion in Argument* (1992), from which the title of this chapter was borrowed. In this work, Walton argues that emotional appeals can be reasonable in selected contexts, in opposition to some traditional views that dismissed emotional arguments as inherently irrational. His study uses Pragma-Dialectical concepts, such as dialogue types and dialectical shifts, to evaluate

B. Konat, *Emotional Appeals in Argumentation*, Argumentation Library 47,
https://doi.org/10.1007/978-3-032-06332-8_2

emotional arguments within a normative framework. Walton's contribution serves as a foundational text for the work presented in this book, as it provides a theoretical framework to understand the nature of emotional appeals. In this book, I aim at extending Walton's ideas into empirical studies of argumentation, exploring how modern theoretical and methodological advances now allow the observation of emotional appeals in natural language contexts. He wrote about his own work:

> the project here is a logical, normative, critical, and linguistic one [. . .] and is meant to have implications for empirical studies of argumentation by social scientists (Walton 1992, p. 29)

Written more than three decades ago, his call for empirical studies of argumentation sets the course for new generations of scholars. As someone who came of age in the field shaped by Walton's insights, I view his statement as a direct invitation to extend his project to the empirical domain. This book aims to fulfill the vision Walton sketched, by integrating theoretical and methodological advancements to analyze emotional appeals in natural language argumentation. Through this work on pathos, I seek to honor Walton's legacy while carving out new directions for understanding how emotions function in argumentation, particularly in real world contexts.

The Place of Emotion in Argument (1992) was written during Walton's stay in the Netherlands in 1989–1990, and it builds directly on the Pragma-Dialectical framework, particularly its emphasis on dialogue types and shifts as central to evaluating the reasonableness of arguments. Following van Eemeren and Grootendorst (1987), Walton looks at argumentation as a context-dependent process governed by rules and dialogue goals. Walton uses these concepts to analyze four so-called emotional fallacies—*ad populum*, *ad misericordiam*, *ad baculum*, and *ad hominem*, arguing that emotional appeals are fallacious only when they cause an implicit, illegitimate shift between dialogue types (Walton 1992, p. 211).

Walton notes that all fallacies analyzed by him contain an element of *ad*, that is, being directed *at* something, towards something—in this case, an audience. He writes:

> These four fallacies all rely on the prejudices in an audience; to stir them up a speaker directs an argument at what he or she takes to be the deeply held emotional commitments of the audience. (Walton 1992)[3–4]

The critical discussion is the model dialogue type in Pragma-Dialectics, designed to resolve differences of opinion through a systematic exchange of arguments and critical questions. Walton claims that fallacious use of arguments *ad* can lead to critical discussion slipping into quarrel, or into persuasion dialogue—a types of dialogue with different rules and different goals. In his view, while changing dialogue types is of course acceptable, the problem lies in the illicitness of the shift. He acknowledges that the introduction of emotional appeals can sometimes lead to illicit shifting of burden of proof—such as in cases where appeal to pity is so strong that makes the hearer feel the need to justify their actions, whereas in fact the speaker should justify theirs. In other words, the fallacious uses of emotional argumentation cover cases where the appeal makes the argument seem so strong that it does not leave room for any doubt. In general, the caveat with emotional appeals is that they always should be viewed as weak, defeasible arguments, which should give way if new knowledge arises.

Walton borrows from Locke the idea that emotional arguments are not always and universally fallacious. For example, he demonstrates how an appeal to pity may be reasonable in contexts where it aligns with the goals of a critical discussion (Walton 1992, p. 109). In this context, he claims that emotional appeals can "be very positive and contribute fundamentally to the goals of critical discussion"(Walton 1992, p. 25). The use of emotional appeal is fallacious when the appeal itself is so strong that critical questions adequate for such an argument are overlooked (1992, p. 261). In Walton's view a good critical discussion is inherently emotional, as it starts with strong partisan argumentation and is directed towards knowledge or truth of a matter. Participants in critical discussion should hold strong opinions about their positions, and emotional argumentation is not only allowed but needed for a good critical discussion. For Walton the only thing that is problematic is illicit shift, or "dark commitments"—a situation where there is either hidden shift in dialogue goals, or there are commitments which are not revealed to other participants. Is it then possible to have rational discussion with emotional argumentation? According to Walton the answer is yes, as he is against the rational—emotional distinction, a view I strongly support in this book.

A key contribution of Walton's work is his concept of rational emotional appeals, which does not accept the traditional dichotomy between reason and emotion. He argues that emotional appeals can be rational when they adhere to the rules and goals of a particular dialogue type. For example, Walton critiques the positivist perspective that views emotional arguments as inherently irrational, stating that "biased argument is not necessarily a fallacious argument" (Walton 1992, p. 126). He provides examples, such as reasonable appeals to pity or even cases where *ad hominem* arguments may be appropriate, such as when questioning the credibility of a sole witness (Walton 1992, p. 106). Walton of course points to the fact that these arguments are defeasible, i.e. they should give way when new knowledge will appear. The value of emotional appeals for Walton lies also in that they can be "good presumptive guides to action or to a provisional conclusion in opinion-based dialogue" (Walton 1992, p. 255). He points out that, even though ideals of critical discussions are worthwhile pursuing, in everyday life we rarely have means and knowledge for fulfilling all the rules of it. Emotional appeals are useful when we have practical but non-conclusive reasons, and when emotional arguments can be used to shift the balance, or to provide interlocutors with provisional defeasible reasoning pointing to a prudential course of action. This is also the area studied in this book—emotional appeals as they are used in everyday persuasive settings, such as televised pre-election debates, academic communication, and competitive debating by students.

Walton claims that "detachment of emotions is not always a good thing in argumentation" (Walton 1992, p. 268)—he emphasizes that emotions provide us with standpoint, with attitude, and they inform us about how we feel towards the discussed issue. This appraisal part of the emotional component mentioned by

Walton can in my opinion be directly translated to psychological theories of emotion and the idea that emotion is a directed reaction to the stimuli (more about it in Part II of this book). This shows how even though Walton did not refer directly to psychological theories of emotions, and he overtly separated himself from them by positioning his work as normative rather than empirical endeavor, still his views are compatible with psychological approaches, as will be evidenced later in this book. Walton's critique of the dichotomy between rationality and emotionality has significant implications for argumentation theory and practice. He condemns the "positivistic dichotomy between facts and sentiments" (Walton 1992, p. 106), arguing that dismissing emotional appeals as non-rational can result in biased judgments. This critique aligns with concerns from feminist approaches to argumentation (elaborated later in this chapter), which emphasize how emotionally charged arguments are often unfairly dismissed, perpetuating systemic prejudice against individuals or groups stereotypically perceived as more emotional. Walton's focus on emotional appeals, different from the simple expression of emotion, also clarifies how speakers try to elicit specific emotions in their audience for rhetorical purposes. His work challenges scholars to move beyond the simplistic view of emotion as a fallacy, reframing it as a critical element of effective argumentation when used appropriately. The research presented here builds on Walton's theoretical foundation by extending his normative framework to empirical analysis. Whereas Walton primarily focuses on evaluating emotional appeals within structured dialogues, this study explores how emotional appeals operate in natural language argumentation. For example, contemporary tools such as sentiment analysis in discourse studies and computational linguistics provide new methods for examining the rhetorical use of emotions. Following Walton's distinction, this study focuses on emotional appeals, deliberate actions aimed at eliciting emotions in the audience, rather than the expression of emotions. This aligns with Walton's methodological insights while addressing how emotional appeals influence real-world argumentation. By empirically analyzing these appeals, this work aims to expand our understanding of their role in both rhetorical effectiveness and critical discussion.

As mentioned, a key part of Walton's framework, following Aristotle, is the focus on emotional appeals, that is, how a speaker tries to elicit emotions in the audience to support their argument. This is different from the expression of emotions, which are often the focus in computational sentiment analysis or some discourse studies. Walton, like Aristotle, is interested in how emotions are deliberately evoked for rhetorical purposes, not just how emotions are expressed by the speaker or writer. Appeals for him are something the speaker does "to the mind of the audience" (Walton 1992, p. 9). If we add inherently pragmatic orientation of Walton, and this relation to "the mind" we can see how modern cognitive pragmatics (as described later in Part I here) fits perfectly into this theoretical model. My work follows this tradition, focusing on how speakers use emotional appeals to influence audiences, rather than analyzing expressed emotions. This distinction is central to the approach taken in this book and will be elaborated in detail in Chap. 3, presenting my interactional model of pathos. Walton's observations on the rational use of emotional appeals, grounded in Pragma-Dialectical principles, remain a cornerstone of contemporary argumentation studies. His critique of the rational-emotional dichotomy and his pragmatic approach to evaluating emotional arguments provide a solid framework for analyzing their role in discourse. Building on this foundation, the present study applies Walton's ideas to natural language contexts, bridging the gap between theoretical and empirical approaches to emotional argumentation. This integration emphasizes the enduring relevance of Walton's work and presents new perspectives on the study of emotional appeals.

2.2 Emotive Meaning and Lexical Choice

Macagno and Walton (2014) developed the ideas of emotions in argumentation, starting from the departure point established by Walton: that emotions do have a place in argumentation theory. Macagno went further and deeper into the practical analysis of how the use of emotions and emotional words influences argumentation, including how it can highlight or hide certain issues in persuasive discourse.

Their book presents multiple converging points and several differences from the approach presented here. An obvious point of convergence is the focus on the role of emotions in argumentation. This is based on the assumption that when we argue to convince someone, when we engage in persuasive argumentation, we are naturally not indifferent to the issue at stake. This means that we feel something about the issue and want the hearer to feel the same emotion, or perhaps a different but equally useful emotion to achieve our goal. Obviously, this is not always the case, as pointed out in Chap. 1 and explored in more depth in Chap. 3. Here, I distinguish between two distinct functions of emotional language in argumentation: one, where it expresses the speaker's own emotional stance; and another, where it aims to elicit specific emotions in the audience. Especially in the second sense, Macagno and Walton's work comes close to mine. They rightly observe:

> Words like 'ignorance,' 'indifference,' 'peace,' or 'terrorism' are emotive in the sense that they are used to elicit emotions or change our evaluation of reality. (Macagno and Walton 2014, p. 2)

In this, they note that the use of certain words in a persuasive context can elicit emotions in the audience and thereby influence the effect of argumentation. This observation is fully consistent with the position advanced in this book, especially to the concept of affective lexicons, described in Part II.

However, Macagno and Walton propose a slightly different view on how this process occurs. They show how the use of emotion-eliciting words can itself be treated as argumentative. When we, as speakers, select certain words, i.e. when we make our lexical choices, we already present ourselves as being in favor of or against the issues at hand. In this respect, I differ, proposing another, but I believe complementary, approach. In my approach, emotion-eliciting words are interwoven into classical argumentative premise-conclusion structures. In my view, the speaker builds the argument at the level of propositions and, through lexical choices, may choose to make an emotional appeal or not. This will become clearer in the presentation of my proposed pathos model in Chap. 3, and its operationalization for lexical analysis in Part III. For now, let us return to Walton's and Macagno's useful observations on how emotion-eliciting words influence the argumentation process.

They first examine how the use of words with positive and negative connotations appeared in definitions in classical writing (p. 12). I find this a particularly useful approach: using this lens to analyze emotional valence in Cicero or Cornificius already shows how we are easily blinded and assume many things to be "neutral" or "rational", when closer analysis reveals the presence of emotions. An example, taken from *Rhetorica ad Herennium* (II, 41, as cited in Macagno and Walton 2014), is: "An informer, in short, is worthy of death; for he is a wicked and dangerous citizen" (Macagno and Walton 2014, p. 2). Which the authors comment on as:

> The purpose of such definitions is not to clarify the meaning of a word, or to describe it; on the contrary, the meaning is altered and extended, but to highlight the negative features of the definiendum, the person classified as 'informer' in this case, supports the further conclusion that he needs to be put to death. (Macagno and Walton 2014, p. 2)

Thus, the authors show how even in classical writing related to court cases and law, speakers created definitions and arguments in a way that aimed, through lexical choices that elicit emotions, at highlighting and hiding certain properties of the discussed issue.

The concept of highlighting and hiding naturally leads us to cognitive linguistics and one of its founding sources: Lakoff and Johnson's *Metaphors We Live By* 1980. The idea that conceptual metaphors are mechanisms of highlighting and hiding means that they make certain aspects of a concept more salient while concealing others. A foundational example offered by Lakoff and Johnson, incidentally related to argumentation, is the metaphor 'ARGUMENT IS WAR', which structures how we think and talk about arguments: we defend positions, attack points, win or lose debates.[1] This metaphor highlights the adversarial, competitive nature of argumentation, but at the same time hides cooperative or exploratory aspects of argument as a form of mutual inquiry or understanding. As Lakoff and Johnson famously suggest, if our culture instead conceptualized argumentation as dancing, a coordinated activity between partners rather than a battle, we might emphasize rhythm, responsiveness, and mutual engagement over conflict and victory.

These concepts of Lakoff and Johnson are in a very good way aligned with Macagno's and Walton's observations on how the selection of emotion-eliciting words hides and highlights certain aspects of argumentation and how it inevitably leads to an evaluation of the issue, in modern or classical discourse. This alignment is rooted in a deep philosophical assumption shared by argumentation theory and cognitive linguistics: that words are not just words but operate at a cognitive level, moving something within the minds of speakers and hearers. Only with this

[1] Following the convention introduced by Lakoff and Johnson, conceptual metaphors are written in small caps and the format X IS Y signals the mapping from a source domain to a target domain.

assumption, often tacitly accepted, can we, as argumentation theorists, meaningfully talk about someone being convinced or changing their opinion. This is why assuming some level of cognitive processes in analyzing persuasion is naturally needed and will be present in the analysis developed throughout this book. However, I believe that a deeper integration of newer findings in cognitive linguistics with argumentation theory holds great potential and deserves separate consideration. For now, let us leave these preliminary notes and return to a more detailed analysis of emotional words, as presented by Walton and Macagno.

Macagno and Walton argue, based on previous work, that emotive, ethical, and evaluative terms are often interrelated and at times conflated. Ethical terms influence attitudes precisely because they function as value judgments. As the authors claim, these judgments classify reality as 'good' or 'bad' and possess a twofold structure: they are grounded in reasons and carry an imperative dimension that influences the interlocutor's interests and future decisions. In this respect, I depart from the view of Macagno and Walton. Although it is difficult to dispute that attributing positive or negative emotional valence to a word amounts to a value judgment—"it is good (to me)" or "it is bad (to me)"—I propose a different perspective, grounded in classical psychological theories of emotion, notably those of Ekman (1999) and Plutchik (1980). I argue for a simplified psychological framework, rooted in evolutionary theory and stimulus-response paradigms, which will be developed in Part II. This perspective, while distinct, still aligns with the conceptual direction of Macagno and Walton. Plutchik defines emotions as biologically driven responses to stimuli oriented around attraction and aversion, thus guiding humans and animals toward desired objects or away from harmful ones. This basic orientation is compatible with more culturally embedded theories linking emotions and values. However, for the sake of conceptual clarity in the pathos proposal in this book, I exclusively adopt the psychological approach. Therefore, I will not engage in detail with the extensive literature on values and ethical argumentation, not because such connections are irrelevant, clearly they are, but because my aim is to lead readers along a coherent path toward the methods for analyzing emotive language, which I present in Part III.

Another important aspect of Walton and Macagno's work that I will integrate into my account is the idea that individual words, regardless of context, can elicit emotional reactions. The authors argue that words activate affective responses through mechanisms grounded in their semantic content, cultural associations, and implicit inferential patterns. Some words, such as 'war' or 'terrorism,' inherently trigger emotions like fear or hostility, thereby shaping attitudes and decisions. Drawing on Stevenson, they distinguish two types of emotive meaning: dependent and independent. Independent emotive meaning resides within the word itself, connotations, experiential associations, and essentialist presuppositions. For example, 'cur' evokes more negative reactions than 'dog' despite referring to the same referent, a phenomenon frequently used in persuasive language (as in 'vagrants' vs. 'homeless'). Dependent emotive meaning, on the other hand, arises from a shared evaluation of a word's referent, shaped by interpretive frames: background knowledge and socially shared values. For example, 'carrion' is perceived negatively because its frame involves undesirable associations such as 'scavenger', 'food' or 'corpse'. Frame theory plays a key role in understanding how emotional language operates in persuasive discourse. However, one issue remains salient: the role of context. While this question lies at the core of pragmatics and it is widely acknowledged that context powerfully shapes the emotional impact of language, it is difficult to account for context adequately with the methods adopted in this book, namely, psycholinguistic experiments and computational approaches. Experiments rely on controlled but artificial environments, which limit ecological validity (although I will present one case study of a so-called 'natural experiment' conducted by my team). Computational methods, as they currently stand, allow for precise categorization of language use, but often at the cost of contextual detail (as will be illustrated in Part III with pathos mining techniques). That said, the development of large language models is beginning to make it possible to incorporate richer context into computational analysis. Future advances in computational pragmatics may offer more reliable ways to capture the pathos in context. However, for now, this limitation must be acknowledged. What I take from Macagno and Walton is the compelling idea that some words possess the power to evoke emotions regardless of context, which is a claim strongly supported by psychological research on affective lexicons, which I will discuss in Part II.

To conclude, this section has traced the conceptual foundations of emotional language in argumentation by engaging closely with the work of Macagno and Walton. Although my own approach differs from theirs in its grounding in psychological rather than evaluative or ethical theory, I fully acknowledge the value of their insights, especially on the emotive potential of individual words and their role in highlighting and hiding aspects of argumentative discourse. The notion that words can function affectively even outside their immediate context aligns with both classical rhetoric and empirical psychology and serves as a central tenet of the pathos model developed in this book. In subsequent parts, I aim to build on this approach with a methodologically tractable model that combines psycholinguistic theory, lexical analysis, and computational tools to better understand how emotions shape persuasive argumentation.

2.3 Other Perspectives

2.3.1 Feminist Approaches: Catherine Hundelby

A look into feminist and queer approaches to argumentation is crucial for my work on pathos in argumentation, because the familiar opposition between "rational" and "emotional" is itself gendered. Western intellectual history routinely codes rationality as masculine and emotion as feminine; as a result, emotional appeals are undervalued precisely when the speaker, or the style, can be read as feminine. Unpacking that bias is essential if we wish to understand how emotion actually works in argumentation.

Douglas Walton's *A Place of Emotion in Argument*(1992), dismantles the empirical basis of the dichotomy. Surveying political speeches and courtroom rhetoric, Walton shows that successful arguers rely on emotion to frame issues, establish stakes, and motivate audiences. Purely "cold" reason is the exception, not the norm. Feminist argumentation theory adds a complementary insight: the dichotomy is not only descriptively false, but also normatively loaded. As Catherine Hundleby observes (2021), labeling a contribution "emotional" often functions as a gatekeeping device: the argument is marked as less objective, less serious, and therefore less worthy of acceptance—all attributes historically projected onto women and other marginalized groups. This explains why the title of this section contains the word "other": it points to the notion of "otherness", a term often used to conceptualize marginalized groups.

The stereotype fails every empirical test. Walton's own corpus consists mainly of male speakers whose arguments succeed because they mobilize emotion. The analyses reported in later parts of this book confirm the pattern: emotional appeals permeate pre-election debates and student discussions irrespective of the arguer's gender. Neuroscientific evidence converges on the same conclusion: affective processing underwrites practical reasoning in every human brain (Damasio 1994). Taken together, these findings suggest that the rational-emotional hierarchy is inseparable from a social hierarchy: masculine over feminine. If we wish to evaluate arguments on their merits, we must dissociate epistemic standards from gendered expectations. This section therefore turns to Hundleby's framework for detecting when credibility judgments are distorted by stereotypes and outlines analytic tools to highlight the positive epistemic role of emotion in argumentation.

Hundelby (2021), in alignment with other feminist philosophers, critically examines the role of emotion within argumentation by pointing to the pervasiveness of what she calls an "Adversary Method" in Euro-American philosophy, which is deeply interwoven with masculine ideals and rational-emotional dichotomy. She notes that common ideals of masculinity and rationality are often equated with aggression, success, and power, while being juxtaposed with emotion, unreason, and other characteristics culturally coded as feminine. This gendered framework heavily influences the disciplinary culture of philosophy. One of the sources she builds this on is Lakoff's and Johnson's insight of "ARGUMENT IS WAR" conceptual metaphor (as described in the previous chapter) underlying the way we speak about arguing. Hundelby claims that this leads to a hostile environment where militaristic language and competitive evaluation foster the eristic goal of defeating opponents and their views. As I will show in this book, emotional appeals not only are present and frequent in various speeches of speakers regardless of their gender, but also effectiveness of such argumentation on hearers is influenced by a myriad of complex factors, among which gender is just one of many. Again, just as rational-emotional dichotomy does not stand in light of empirical evidence, a complementary one—male-female—also cannot be held.

Hundleby points out that this entrenched adversarial paradigm, which she identifies in critical thinking textbooks, often reinforces disciplinary biases by promoting a narrow discourse focused on decisive refutation. Such a context can disadvantage women, as their contributions, if presented in styles perceived as feminine, may not be noticed or deemed competent. Furthermore, when women do adopt aggressive, "masculine" discursive styles, they often face backlash, being criticized as selfish, cold, and mean, which can ultimately weaken their authority and ability to participate effectively in argumentation. She also acknowledges that gendered politeness norms can further subordinate women in argumentative contexts, as deferential styles associated with femininity may imply powerlessness, while masculine norms facilitate shared competition.

Crucially, Hundleby, does not abandon the role of anger, an emotion that is just as useful as others. She contends that arguing provides an avenue for reasoners to express and articulate their underlying reasons, thereby revealing aspects of reasoning that might otherwise remain unconscious. This indicates that for Hundleby, emotions are not simply impediments to reason but can be integral to a more comprehensive understanding and articulation of arguments, and just like Walton, she is challenging the traditional opposition between the rational and the emotional in argumentation. Her work also extends to identifying how implicit biases reinforce these gendered norms, proposing the 'status quo fallacy' to describe the assumption of desirability of stereotypical qualities associated with socially authoritative groups, such as men.

To summarize, what is crucial from Hundelby's work to the readers of this book is the identification of the unconscious bias we might all have, regardless of our own gender, that identifies emotional with the culturally entrenched idea of "female" and the lack of emotions with a similarly constructed idea of 'male'. To begin our journey towards understanding emotional appeal, a realization of this bias is a necessary step.

2.3.2 *Queerness and Emotion: M.A.Gilbert*

Another interesting perspective challenging cultural dichotomy comes from Michael A. Gilbert, a scholar with immense input into informal logic and critical thinking teaching. What is interesting in this context is that they are also self-identifying as a queer person, publishing also under the name of Miqqa A. Gilbert, and self-describing as a cross-dresser or a transgender person. Just as emotional is perceived as the "other" to rational, female is perceived as the "other" to male, so is queerness perceived as the "other" to the rest of the population. Starting from this perspective, Gilbert in their later work explicitly contextualizes their classical works, including influential work on emotional argumentation, in their own lived experience of queerness.

M. A. Gilbert has devoted a substantial portion of their academic career to publishing analyses and manifestos that highlight the significant role of various non-logical elements in argumentation and decision-making. Given the aim of this book, this section will focus on their contributions regarding the role of emotions in argumentation. The first step in Gilbert's approach towards the notion that arguments can encompass more than just logical modes of communication, paving the way for further exploration into the role of emotion in argumentation, is the concept of multi-modal argumentation (Gilbert (1994) with the revised version 2022). They initially stated "that I believed in the importance of reasoning, especially as an alternative to violence" (Gilbert 2022). Their work on the now-renowned book *How to Win an Argument*, which was originally designed for adult education, served as a catalyst for their growing interest in real-life arguments (Gilbert 1978). Another reason for Gilbert's journey was that they took a break from academic life and for a couple of years was writing fiction. In their own words, this led them to think that what is often studied as examples in argumentation textbooks does not resemble how real people actually argue. I have to pause here and say that I admire this approach, and I can see how a fiction writer's need for a keen ear for natural language dialogues can be valuable in studying real-life argumentation and how people persuade one another. As I will continue to argue and hopefully provide evidence in this book, for us scholars to understand why people can be convinced on social media to do seemingly irrational things (such as drinking bleach or spreading fake news), we need to first examine deeper the everyday dialogue to identify what serves as a persuasive factor. If we listen closely to natural language interchanges, we could see that they contain much more than what could academically be considered as "pure reason". This is the conclusion that M. A. Gilbert reached when they returned to academia after some years of writing fiction: There is more to real-life argumentation than a mere premise-conclusion structure (2022). This led to their vision of multi-modal argument: an argument that occurs in multiple modes of communication. They distinguish four modes in which an argument occurs in the real world:

- *logical*: assists in moving from a message to a conclusion in a reasoned and patterned way
- *emotional*: emotions being expressed in or by an argument
- *visceral*: all aspects of a message or an argument that are physical or environmental
- *kisceral*: intuitive, mystical, religious, or revelatory

Gilbert claims that a logical mode is present in some way in all arguments (Gilbert 2022). This holds even for what Perelman and Olbrechts-Tyteca describe as quasi-logical arguments (1969). Based on my reading of Gilbert, and in line with the stance I adopt from Inference Anchoring Theory (IAT, Budzynska and Reed 2011), a key condition for recognizing something as an argument is that it presents a clear premise-conclusion structure. This structure allows a listener (or, in contexts like those studied in this book, an analyst or annotator) to identify the argument as such. Even when the reasoning is flawed or informal, the presence of this structure makes the communication recognizable as argumentative.

The emotional mode in Gilbert's approach focuses on expressed emotion; they point out that often the emotional message of an argument can be so strong that it can eclipse what has actually been said. Although I agree with this stance regarding the importance of emotion in argumentation, in this book I focus less on expressed emotions and more on how emotions can be elicited in the hearer for rhetorical gain. As mentioned, in this regard, I am closer to the classical Aristotelian concept of pathos than to modern approaches to expressed emotion. This does not mean, of course, that I negate the influence of expressed emotions on the persuasive power of an argument. On the contrary, in the theoretical approach presented here, emotions expressed by the speaker can influence the hearer's emotional state and their receptiveness to the argument being made. For example, my expression of anger can lead to you feeling fear, and therefore an argument can further build on that fear. However, what I do not share with Gilbert is the focus on expressed emotions; in my model, they are just one of many means of eliciting emotions in the hearer. The combination of logical and emotional elements in natural argumentation is something that is present in both Gilbert's approach and the model presented in this book.

Two other modes proposed by Gilbert are *visceral* and *kisceral*.The first refers to both a widely understood context (power relations or physical configurations) and actual physical movements accompanying the argumentation (example included in Gilbert (2022) is "[[Robert touches Marcia's hand]]"). This mode brings of course to mind pragmatics of argumentation, a discipline that flourished in recent years and that is described in the next chapter of this book. However, pragmatics itself is not the central topic of this work, it is important to remember when analyzing natural language argumentation that this is something that occurs—like any linguistic event—within certain context of real-life world. The name of the last mode (*kisceral*) Gilbert has based on the Japanese word "Ki"- "energy". This mode relates to intuitive or sacred cognition. It should be noted that while the *visceral* and *kisceral* modes offer intriguing possibilities for understanding the complexity of real-world argumentation, they fall outside the scope of this book's focus on the intersection of *logical* and *emotional* modes.

M. A. Gilbert embarked on their exploration of real-life argumentation in 1978, and by 1994 they were laying down the foundations for multi-modal argumentation. However, as they point out in the 2022 paper, this important work has been acknowledged but not fully integrated into the dominant methodologies for analyzing arguments (Gilbert 2022). They ascribe it either to the rules of Pragma-Dialectics (van Eemeren and Grootendorst 1987) or to the challenges inherent in diagramming emotions that are not directly expressed. Setting prescriptive approaches aside, this book introduces a method that, I believe, accommodates the inclusion of emotional content within diagrams created using the IAT methodology. Exploiting computational tools that allow for multiple layers of text annotation, we can capture the emotional valence of language used in the premises or conclusions of natural language arguments. By employing psychological theories and methods for gathering emotional assessments of language samples, we can obtain measures that, while perhaps not objective, are at least intersubjective. These methods and results are discussed in Parts II and III of this book. For now, let us go into more detail into Gilbert's proposed contributions to the main fields of study related to natural language argumentation.

When analyzing the relationship between emotions and informal logic, Gilbert points out that despite integration with other disciplines such as social psychology and an increasing body of work dealing with emotional arguments, in informal logic "the tide has not changed dramatically" (2004). The division between "rational" (linear, discursive, based on empirical statements) and "non-rational" arguments is still visible. This is despite the fact that even the founding work of modern informal logic correctly assumes that almost no argument found in natural language can be considered fully formal (Perelman and Olbrechts-Tyteca 1969). In this context, Gilbert claims,"the utilization of emotion in arguments, in part or in whole, is perfectly rational"(2004).

They then focus on the expressed emotion and how it can be inferred from the tone of voice or other contextual clues. Because this is the case, they claim that an informal logician, when analyzing an argument, should be obliged to look into this context, which for them consists of the four modes described earlier in this section. They claim that, following the tradition of being at least cautious with emotions, presented in argumentation textbooks (Copi 1963; Govier 2001), informal logicians tend to abstract from emotional content. Since the emotional component is found in almost all human interactions, this is not an adequate approach towards analyzing real-world argumentation. One of the reasons for such reserve is, claims Gilbert, the fuzzy, inclined-to-bias, and misinterpretation nature of the emotional component of argumentation.

Gilbert moves on the concept of assessing the argument: What would it mean for an emotional argument to be acceptable within argumentation theory? Gilbert uses the following example to show the differences between truth conditions related to claims such as A–D. (2002):

A. I am six feet tall.
B. Colette is six feet tall.
C. I am angry.
D. Andrew is angry.

Gilbert rightly claims that assertions like A. and B. have different truth conditions than C. and D. and therefore can be accepted under different conditions as argument premises.

In this paper, Gilbert (2002) uses hypothetical dialogues featuring interlocutors discussing academic scenarios such as the role of student enthusiasm in course selection or the impact of personal sadness on promotion decisions. Although these examples are valuable for theoretical considerations, it is important to note that contemporary computational tools offer us a different perspective. With the advent of data analytics and the ability to collect real-world conversations from social media or transcribed dialogues, we observe that people rarely articulate emotions in such an explicit way. Therefore, the work presented in this book aims to build on Gilbert's foundation by incorporating modern methodological advancements.

In summary, the integration of feminist and queer perspectives into argumentation theory not only broadens the conceptual framework for understanding but also exposes the sociocultural assumptions that have long shaped what counts as rational argument. Scholars such as Catherine Hundleby and M.A. Gilbert demonstrate that the dichotomy between reason and emotion is not just analytically flawed but also ideologically charged, reproducing hierarchies of gender, sexuality, and epistemic authority. Their work emphasizes the importance of attending to the full range of communicative modes and contextual factors in real-world argumentation. These works complement and extend Walton's normative framework, showing how emotional appeals operate not just as rhetorical tools but as markers of inclusion or exclusion in epistemic communities. The empirical studies presented in the following chapters aim to carry forward this interdisciplinary trajectory, investigating emotional appeals in authentic discourse while remaining attentive to their normative, social, and political dimensions.

References

Budzynska K, Reed C (2011) Speech acts of argumentation: Inference anchors and peripheral cues in dialogue. In: Computational models of natural argument: papers from the 2011 AAAI Workshop, AAAI Press, pp 3–10

Copi IM (1963) Introduction to Logic, 2nd edn. Macmillan, New York

Damasio AR (1994) Descartes' Error: Emotion, Reason, and the Human Brain. G.P. Putnam, New York

van Eemeren F, Grootendorst R (1987) Fallacies in pragma-dialectic perspective. Argumentation 1(3):283–302

Ekman P (1999) Basic emotions. In: Dalgleish T, Power M (eds) Handbook of Cognition and Emotion, Wiley Online Library, pp 45–60

Gilbert MA (1978) How to Win an Argument. McGraw-Hill, New York

Gilbert MA (1994) Multi-modal argumentation. Philosophy of the Social Sciences 24(2):159–177

Gilbert MA (2002) Effing the ineffable: The logocentric fallacy in argumentation. Argumentation 16(1):21–32

Gilbert MA (2004) Emotion, argumentation and informal logic. Informal Logic 24(3)

Gilbert MA (2022) Multi-modal 2020: Multi-modal argumentation 30 years later. Informal Logic 42(3)

Govier T (2001) A Practical Study of Argument, 5th edn. Wadsworth Thomson Learning, Belmont, CA

Hundelby C (2021) Feminist perspectives on argumentation. Stanford Encyclopedia of Philosophy https://plato.stanford.edu/entries/feminism-argumentation/, first published Thu Feb 18, 2021

Lakoff G, Johnson M (1980) Metaphors We Live By. University of Chicago Press, Chicago

Macagno F, Walton D (2014) Emotive Language in Argumentation. Cambridge University Press, https://doi.org/10.1017/CBO9781139565776

Perelman C, Olbrechts-Tyteca L (1969) The New Rhetoric: A Treatise on Argumentation. University of Notre Dame Press

Plutchik R (1980) Chapter 1 - a general psychoevolutionary theory of emotion. In: Plutchik R, Kellerman H (eds) Theories of Emotion, Academic Press, pp 3–33, https://doi.org/10.1016/B978-0-12-558701-3.50007-7, https://www.sciencedirect.com/science/article/pii/B9780125587013500077

Walton D (1992) The Place of Emotion in Argument. The Pennsylvania State University Press, University Park, PA

Chapter 3
Towards the Model of Interactional Pathos

Abstract This chapter introduces the Model of Interactional Pathos in Argumentation (MIPA), a new theoretical framework for analyzing emotional appeals as intentional, context-sensitive speech acts. Drawing on cognitive pragmatics, particularly Relevance Theory, as well as contemporary developments in argumentation theory, the chapter argues that pathos should be understood as an interactional event occurring between speaker and audience. Emotional appeals are conceptualized as pragmatic acts that seek to increase the perceived strength of an argument by eliciting emotions in the hearer. The chapter further explores the linguistic realization of pathos through pathos-related argument schemes and discusses how these can be operationalized using both manual and computational methods. The theory of speech acts provides the foundation for this model, distinguishing between the intention of the speaker, the linguistic realization and the response of the audience. Emphasizing the integration of emotion, argument, and pragmatics, the model provides foundations for empirical research on emotional appeals in natural language discourse. The chapter concludes with an annotation study of political debates, demonstrating how MIPA supports systematic analysis of pathos in real-world argumentation.

Keywords Interactional model of pathos · Pragmatic argumentation · Emotional appeal · Relevance Theory · Speech act theory · Pathotic argument schemes

3.1 Cognitive Pragmatics and Relevance Theory

Cognitive pragmatics is primarily concerned with the psychological processes involved in language comprehension and communication. One of the seminal frameworks in this domain is Relevance Theory, developed by Sperber and Wilson (1996). Unlike traditional semantic theories that rely solely on decoding sentences, Relevance Theory places a strong emphasis on inferential processes for understanding communicative intentions. Thus, the core of Relevance Theory is understanding communication as an inferential process. The listener uses various cues, not just the semantic content, to infer the speaker's intended message. The theory postulates

B. Konat, *Emotional Appeals in Argumentation*, Argumentation Library 47,
https://doi.org/10.1007/978-3-032-06332-8_3

that this inferential process is part of a broader cognitive phenomenon known as "relevance", which is geared towards efficient information processing. This concept of relevance is not specific to language but is a fundamental aspect of cognitive systems in both humans and other animals. It serves as a mechanism for focusing attention on the most significant environmental signals, helping to survive and learn. The theory is anchored in two main principles (Huang 2020):

1. *First Principle: Maximization of relevance*: Human cognition aims to achieve the maximum informational benefit for the least cognitive effort. This principle is not just confined to communication, but is a general feature of human cognition.
2. *Second Principle: Presumption of optimal relevance*: In any act of explicit communication, the speaker provides information that is both sufficiently relevant and as easy to process as possible, given her abilities and preferences.

In this framework, the concept of implicatures, initially introduced by Grice (1975), undergoes significant elaboration. While Grice's implicatures arise from conversational maxims, Relevance Theory incorporates them into an inferential framework guided by the principles of relevance. Both "explicatures," derived primarily from linguistic content, and "implicatures," inferred from context, function with the aim of maximizing relevance, defined as the ratio of information gain to processing effort.

Building on Grice's concept of "Maxim of Relevance", Sperber and Wilson develop it using the cognitive foundation of maximization of relevance. They propose that each act of ostensive communication includes the presumption of optimal relevance, which consists of two components:

1. *Presumption of optimal relevance*:
   (a) The set of assumptions *I* which the communicator intends to make manifest to the addressee is relevant enough to make it worth the addressee's while to process the ostensive stimulus.
   (b) The ostensive stimulus is the most relevant one the communicator could have used to communicate *I* (Sperber and Wilson 1996, p. 158).
2. *Principle of relevance*: Every act of ostensive communication communicates a presumption of its own optimal relevance (Sperber and Wilson 1996, p. 158).

Sperber and Wilson also recognize the limits of this model. Not all forms of communication fall under the category of ostensive-inferential communication; certain types of coded signals may operate differently. Moreover, communicators may deliberately exploit the presumption of relevance for manipulative ends. As they note:

> The principle of relevance does not say that communicators necessarily produce optimally relevant stimuli; it says that they necessarily intend the addressee to believe that they do. Even bores manifestly intend their audience to believe that they are worth listening to (Sperber and Wilson 1996, p. 158).

In summary, Relevance Theory builds on the assumption that human communication works just like other cognitive systems: it is geared towards obtaining valuable ("relevant") information, i.e., the type of information that will allow for the optimal action. At the same time, relevant means that to obtain this information, the hearer will devote as much effort as needed—but not more. These two assumptions allow the authors to propose that every act of ostensive communication conveys not only its message, but also a presumption that it is relevant to the hearer. In other words, when I tell you something, I am doing this because I believe this is relevant to you. This concept is extensively developed in the work by Oswald (2010), devoted to pragmatic views on manipulation. This is also where emotions come into play.

3.2 Pragmatics of Emotion

Taking Relevance Theory as a starting point, Steve Oswald explores how emotions can function within manipulative communication, especially when speakers exploit relevance assumptions. In Oswald's work, uncooperative and manipulative communication is characterized by the speaker's effort to make use of the fallibility of human perception. In analogy to the illusionist, who is waving one hand in front of the audience to shift the attention from the other hand doing the trick, the manipulative speaker is using the assumption of relevance to make some information salient and obscure the other information (Oswald 2010, pp. 335–336). Using the cognitive relevance principle, the manipulator shifts hearer's attention to put external constraints on both explicatures and possible implicatures of what has been said. Oswald treats emotion as one of the "constraints" in his proposed model of Contextual Selection Constraints (Oswald 2010, p. 345). In this model, manipulative communication is examined through the lens of cognitive pragmatics. The essence of manipulation lies in how the manipulator emphasizes certain elements of the context while downplaying others to guide the hearer towards an interpretation favorable to the manipulator. Throughout this process, the speaker operates under the assumption that the principle of relevance is being followed, that their interpretation is the most relevant, given the cognitive effort expended. Within the model proposed by Oswald, emotions play an important role: the emotional state of the hearer can influence which information from the communicative stimuli will be accepted as relevant. As Oswald postulates, the listener can, from the set of contextual assumptions, select those which are consistent with their emotional state. These initial ideas were later developed by Oswald and colleagues studying the impact that affective science can have on the developments of linguistic pragmatics.

In the work titled *Relevance and Emotion* (Wharton et al. 2021), a group of European authors representing both affective sciences and linguistic and pragmatic studies advocate for a comparative analysis between these fields. Their aim is to better define "emotional relevance" and present the need for cross-disciplinary engagement. In this position paper, the authors emphasize that while affective scientists are deepening our understanding of emotional processes, pragmaticians tend to include emotions as part of the context without looking into the underlying psychological mechanisms. In contrast, communicative aspects of emotion, such as elicitation and expression, subjects extensively studied within the field of pragmatics, are often neglected by affective scientists. According to the authors, this state of affairs calls for closer cooperation between the disciplines. What is their specific proposal?

The authors point to the overlap between these two areas of study—the shared concept of "relevance". Although relevance has been at the center of modern pragmatics, as presented earlier in this chapter, the notion of relevance, while differently used, is present in affective science as well (Wharton et al. 2021). The authors acknowledge that regardless of the theoretical model, the concept of relevance or a similar notion like significance, importance, or value is crucial for emotion elicitation. Affective scientists agree that for an object to elicit an emotion, it must be relevant to the person experiencing the emotion. Specifically, terms such as "refer to relevance", "goal relevance", or "motivational relevance" are often used to describe this phenomenon. The authors explore appraisal theory of emotion, as the one where the assessment of the stimuli as relevant is necessary for the later appraisal—deciding which action should be taken based on emotion elicited by these stimuli. In general, the authors claim that the exact understanding of the concept of relevance in affective sciences and linguistic pragmatics is different; they do, however, have overlapping elements. For this, the authors propose a more general concept of "goal relevance", which provides a broader context that encapsulates both types of relevance. As "In the same way the search for relevance is a search that underpins human cognition, it is also the search that underpins human emotion." (Wharton et al. 2021, 267), the authors call for further integration between the two disciplines. This goal is achieved in a subsequent book coauthored by two of the authors of the paper described above: Tim Wharton and Louis de Saussure.

Pragmatics and Emotion (Wharton and de Saussure 2023) offers a comprehensive exploration of the relationship between the communicative context and emotions. The core premise of Wharton and de Saussure's work aligns closely with the stance proposed in this book, namely that the emotional and the rational, or emotion and cognition, should not be separated in theoretical frameworks. The authors contest the traditional dichotomy between "the head" and "the heart", arguing that emotions are central to human communication and must be addressed in any pragmatic theory. Despite these conceptual similarities, their approach is focused on a different aspect from the one explored in this book. Specifically, their work is oriented towards building a pragmatic theory, a theory of utterance interpretation, emphasizing emotional or expressive meaning. In contrast, my work centers on argumentation and reasoning, and the role of emotional appeals in persuasive communication. As my material comes from natural language argumentation, i.e. arguments occurring in context, several ideas presented by Wharton and de Saussure are worth reviewing here.

Wharton and de Saussure identify several challenges associated with using classical pragmatic concepts to describe linguistic expressions of emotions. One central issue is the conflict between description and expression; whereas many phenomena can be readily described using language, emotions resist such straightforward description. Another challenge centers around the notion of propositions and the associated problem of ineffability, i.e. the limitations of how language can express the complexity of emotional states. Traditionally, propositions should be interpretable in terms of truth conditions, which means that they can be true or false. However, if expressions of emotions are ineffable and thus cannot fully convey what an individual feels, they cannot be treated within a truth-conditional framework, as they are not subject to verification.

This conceptualization of the challenges by Wharton and de Saussure points to the distinction between capturing the problem of emotion in pragmatics and in argumentation theory (especially one focused on natural language argumentation). While it is generally accepted that expressing full emotional range in language is difficult and nearly impossible, and at the same time that emotional meaning is often non-propositional and falls outside the scope of truth conditions, these challenges are less problematic in the study of natural language argumentation. As shown in real-life examples through this book, in persuasive speech speakers rarely use pure expressives of the type "I am sad.". Moreover, for persuasive communication, inducing emotions in the audience is arguably even more crucial than expressing them. Pathos appears often next to logos or even intertwined in premises or conclusions with the use of emotive words in seemingly "objective" argument. Often such words are hidden inside the structure of the argument itself. All of this suggests that while pragmatic approaches to emotions are useful for building interactive model of pathos, they explore a slightly different territory, leaving room for a different approach—focusing on interactive nature of emotional appeals.

3.3 Pragmatics of Argumentation

3.3.1 Intersection of Pragmatics and Argumentation

The previous section examined how emotional meaning functions in context, focusing on the relationship between pragmatics and emotion. In this section, we shift attention to a different but complementary pairing: the relationship between pragmatics and argumentation. Although the earlier discussion centered on how emotion operates in communicative practice, the present focus is on how argumentation is interwoven with pragmatic processes. This distinction is crucial for developing the concept of pathos advanced here, which requires bringing together all three dimensions: pragmatics, emotion, and argumentation . As a starting point, I consider existing work on the pragmatic grounding of argumentative activity.

In the introduction to the special issue of *Languages* journal on this topic, Oswald (2023) introduces the idea of a pragmatic-argumentation "interface". Oswald emphasizes that argumentation, being a communicative activity predominantly verbal in nature, naturally incorporates the study of verbal resources and thus all areas of linguistics, particularly pragmatics, are relevant to its study. This interface is significant because both fields share objects of study, such as communicative practices (argumentation being a subset of these) and theoretical notions like inference, commitment, and meaning. Pragmatics offers a set of descriptive, normative, and explanatory tools that are easily applicable to argumentation. This approach is clearly in line with the work presented here. As will be demonstrated in the later parts of this book, tools from linguistics (including pragmatics, psycholinguistics, and computational linguistics) provide a valuable toolbox for studying natural language argumentation. Pragmatics, being one of the most philosophically grounded branches of linguistics, serves in this book as both a foundation and a pathway for the interactional model of pathos.

Among the studies exploring the pragmatic-argumentation interface collected by Oswald, many demonstrate how pragmatic input can improve our understanding of rhetorical effects, including pathos, through both experimental and corpus-based approaches. de Oliveira Fernandes and Oswald (2023) explore the impact of different types of implicit meaning on rhetorical effects, particularly those influencing audiences, including the use of pathos. They propose that rhetorical effects can be understood as perlocutionary effects, a concept rooted in speech act theory. I will build on this claim in later chapters here, when presenting the interpretation of pathos model in speech acts theory. This is built on the assumption that any model of pathos should naturally include the consequences that the comprehension of a message has on conversational participants.

Hautli-Janisz et al. (2023) investigate the role of conventional implicatures in argumentation, starting from the premise that much argumentative reality is implicitly conveyed and serves numerous purposes, from structuring inferences to persuading audiences. Unlike enthymemes, conventional implicatures are anchored in the linguistic surface and yield largely predictable discourse contributions, making them more amenable to systematic study, including via corpus and computational methods. They show how these implicatures can contribute to triggering, composing, and undermining arguments, not just at the textual level but also at the ethotic level, influencing the speaker's image. While their paper focuses on logical inference and ethos, it demonstrates how implicit linguistic elements, discoverable through corpus analysis, can exert rhetorical force and influence argumentative dynamics. In the book presented here, I purposefully do not engage with ethos; however, I believe the work of Hautli-Janisz and colleagues presents a useful analogy of how pathos, just as ethos is studied there, can be studied with the use of solid theoretical grounding in pragmatics, supported by modern corpus approaches.

Tindale (2022) further complicates the picture by exploring Grice's theory of utterer meaning and the challenges of successful communication, particularly in cross-cultural argumentation. According to Grice, successful communication of meaning hinges on the hearer's recognition of the speaker's intention to mean something. This requires speakers to design messages in terms that will be understood, often presupposing a shared common background. Tindale points to the fact that the absence of sufficient contextual conditions, such as shared beliefs and norms, makes this recognition difficult, leading to misunderstanding. The mechanism appears to differ slightly in the case of pathos, the emotional appeal, since the intended emotional effect (a perlocutionary outcome) depends on the audience's appropriate response to the speaker's communicative intentions and to the cultural meanings embedded in the message. If the audience's psychological context does not align with the speaker's, the emotional appeal may fail or be misconstrued. Tindale illustrates this with the example of a dismissive gesture in a parliamentary setting, which, though strictly nonverbal, constituted a meaningful argumentative act with clear cultural currency, interpreted as racist by the audience sharing that cultural background. This is a strong illustration of how, for pathos to be effective, the emotional cues, whether verbal or nonverbal, must be recognized within the audience's specific cultural and contextual framework.

In essence, I fully agree with Oswald that the pragmatic-argumentation interface provides a valuable framework for analyzing pathos. Experimental pragmatics can empirically test how different implicit meanings trigger emotional and other rhetorical effects, while corpus studies can reveal the linguistic mechanisms of implicit communication in real-world arguments. Both methodologies will be presented in Parts II and III of this book. Simultaneously, understanding utterer meaning and cultural contexts highlights the interpretive challenges audiences face, emphasizing that the success of emotional appeal depends heavily on the audience's ability to recognize and interpret intentions within shared cultural knowledge or the potential for misunderstanding when such common ground is absent. This perspective is central to the proposed model of pathos, which necessarily includes both sides: the speaker and the audience.

To further explore the role of pragmatics in argumentation, the following sections will examine two widely influential theories that have generated numerous empirical studies and that conceptualize argumentation within a pragmatic framework: Pragma-Dialectics and Inference Anchoring Theory.

3.3.2 *Pragma-Dialectics*

As the model of pathos proposed in this book is interactive in nature, let us now look at one of the most influential theories, which included the pragmatic aspect in argumentation . Pragma-Dialectics (PD), introduced by van Eemeren and Grootendorst in 1984, marked a significant shift toward an interactional and pragmatic view of argumentation. Rather than treating arguments as isolated logical products, PD conceptualizes argumentation as a communicative activity, a complex speech act performed in natural language, aimed at achieving a reasonable resolution of a difference of opinion. Building on speech act theory and Gricean pragmatics, van Eemeren and Grootendorst analyze argumentative moves as illocutionary acts exchanged between interlocutors. One useful element of PD is its dual nature: it integrates the descriptive aspects of actual language use with a normative commitment to critical reasonableness.

PD thus introduces a functional, interaction-oriented perspective into argumentation theory, viewing argumentation as an organized dialogue directed toward resolving disagreements based on their merits. One of the best-known elements of PD is an ideal 'critical discussion' composed of several stages and governed by ten discussion rules. In examining real-life argumentation, the model recommends reconstructing exchanges into an "analytical overview" that maps the issues under discussion, the standpoints and responses of each party, the explicit and implicit premises and the structure of the argumentation. Starting from this overview, the analyst can evaluate the quality of the argument by checking for logical consistency and pragmatic coherence, assessing the acceptability of premises, the potential validity of inferences, the correct use of argumentation schemes, and the absence of rule violations. In sum, PD, in its dual nature, offers both a structured framework for empirically analyzing how arguers perform speech-act-based moves in discourse and also normative tools for critically evaluating those moves in light of rational standards.

In the *Handbook of Argumentation Theory* (van Eemeren et al. 2014), emotions are not prominently discussed; however, they are mentioned as part of the presentation of selected theories. When introducing the concept of classical rhetoric, the authors include Aristotle's means of persuasion, including pathos. They refer to how, according to Aristotle, persuasion can be achieved by steering the audience's emotions or "putting the audience into a certain frame of mind". A detailed figure in the handbook elaborates on Aristotle's modes of persuasion. In particular, emotions such as anger are mentioned but not explored in depth, leaving room for further study on the topic of emotions within argumentation theory (van Eemeren et al. 2014, pp. 117–119). The book also briefly defines pathotic fallacy as "playing on the sentiments of the audience" (van Eemeren et al. 2014, p. 168). It mentions fallacies such as the argumentum *ad baculum* or the argument of force, which often rely on emotional appeal. Examples of fallacies related to pathos, such as argumentum *ad misericordiam* or the pity argument, are also listed. These fallacies, including those related to pathos, are categorized as violations of the rules of critical discussion (van Eemeren et al. 2014, p. 546). So, if we take the Handbook version of PD as a standard, we can observe a stark contrast in how emotional appeals are treated in PD and in my proposal outlined here in this book. It appears that the core PD framework tends towards the concept of pathotic fallacy. However, there are certain novel developments and additions to PD that show that emotions can also have a place within this theory.

In my opinion, recent developments within the pragma-dialectical tradition have extended its analytical and evaluative apparatus to accommodate greater empirical and communicative complexity. Jean Wagemans's Periodic Table of Arguments (2016, 2023) refines the structural taxonomy of argument schemes, offering a systematic matrix for identifying and classifying arguments based on their inferential structure and topical content. This analytical precision complements efforts to render PD more empirically tractable, as seen in Martin Hinton's Critical Analysis of Argumentative Patterns in Natural Argumentation (CAPNA; 2021), which builds on pragma-dialectical insights to account for recurring patterns in real-life argumentative discourse. At the same time, scholars such as Menno Reijven have pushed the framework toward a closer engagement with conversation analysis, emphasizing the sequential and situated dynamics of argumentative exchanges in everyday interaction (Reijven 2024). These directions preserve the normative orientation of PD while opening it to richer descriptions of how argumentative practices unfold across diverse communicative settings.

There are two other relevant commentaries and developments in PD that I would like to highlight here. The first is Walton's *The Place of Emotion in Argument* (1992), which has been in detail discussed earlier and is intended to serve as a foundational work for this monograph. As mentioned, *The Place of Emotion in Argument*, as the author himself states, was written mainly during his research stay in the Netherlands. The book is openly rooted in PD and clearly shows its influence throughout. Walton aims to develop the PD with two of his key strengths: the analysis of real-life argumentation, particularly in legal and political contexts; and second, the integration of emotional aspects into argumentative analysis. This demonstrates again how introducing the pragmatic and interactive dimension into argumentation theory creates space for the inclusion of emotional components, which naturally and frequently occur in persuasive everyday argumentation. Although PD itself does not focus extensively on emotions, certain developments within and around it point in this direction. I believe that the presence of PD within argumentation theory has been a necessary step toward enabling the inclusion of more interaction-oriented features (such as emotional appeal) into the study of argumentation.

The second important proposal for the development of PD comes from M. A. Gilbert, whose work was already presented in the previous sections. In the paper with the evocative title *Let's talk!* (2005) they propose that the PD model can make space for human emotion within its framework. They posit that no argument, disagreement, or even communication exists without at least a minimal emotional component. Gilbert argues that the PD model is susceptible to reinterpretation in emotional terms by using its core structure as a guide to understanding emotional communication, rather than simply adding emotion and stirring. They demonstrate that three of the four foundations of PD: externalization, functionalization, and socialization; can be straightforwardly adapted to apply to emotional content.

Gilbert supports the idea that the PD model could contain an emotional element through several key arguments. They emphasize that the vast majority of human communication that involves disagreement includes emotional commitment. Crucially, emotions explain why individuals might hold onto an untenable position or defend an indefensible view; understanding a proponent's emotional appeal is often needed to comprehend why a position is maintained. They argue that in many arguments the true issues are not the initial discursive points but the underlying feelings of the proponent. Gilbert views the emotional and logical components as integrated, often so thoroughly that their separation is difficult or impossible. He also points out that exploring why someone holds a position, including their goals and objectives, is not a logical fallacy but a means to open communication and reach agreement, suggesting that emotions can be the most central items considered in a dispute. We can see here a parallel with Walton's view presented earlier, referring to the normative aspect of PD: that emotional elements in the argument do not necessarily mean that the argument is fallacious. What is most interesting for me is that Gilbert suggests that the resolution of emotional arguments aligns with a liberal interpretation of PD's resolution-centered system, where the participants themselves are the ultimate judges of settlement. This is useful as it shows how the inclusion of emotional elements into a theory of argument that already contains a pragmatic component can be seamless and productive. Although Gilbert's proposal was not fully included in the core of PD, it still provides a useful illustration of this integration process.

In summary, while the standard version of PD does not engage extensively with the emotional component, its analytical orientation has opened the door to more empirical approaches and commentaries that acknowledge the frequency of emotional elements in real-life interaction. As a result, even within PD, emotions can be accommodated through these subsequent developments.

3.3.3 Inference Anchoring Theory

Inference Anchoring Theory (IAT; Budzynska and Reed 2011a,b) is a framework for modeling natural language argumentation that focuses on the interactive, dialogical nature of reasoning. Rooted in the traditions of speech act theory (Austin 1975; Searle 1979) and the informal logic diagramming tradition (Toulmin 1958; Reed et al. 2007), IAT conceptualizes arguments as socially situated acts, anchored in conversational turns and illocutionary intentions. Rather than treating arguments as abstract logical forms, IAT posits that they appear in response to dialogical demands when the claim of a speaker is challenged, prompting a justification. Thus, the structure "*p* because *q*" does not exist in isolation; it is constituted through dialogic engagement. It answers the question "Why *p*?" with "Because *q*", illustrating how a speaker provides a reason for their claim (whether the original question is posed explicitly or implied). This perspective builds on a view of language as action: arguing is a pragmatic response embedded in social interaction.

To capture the structure of dialogical arguments, IAT defines three basic propositional relations: inference, conflict, and rephrase. Inference denotes the intentional connection between premise and conclusion within an argument, modeled as the act of arguing and anchored in dialogical transitions. Conflict captures oppositional moves where one proposition directly attacks another, often in the form of rebuttals or rejections. Rephrase represents rhetorical reformulations or elaborations that preserve propositional content but modify its presentation. These relations are complemented by illocutionary connections (such as asserting, challenging, or questioning) that link speech acts to propositional content.

IAT has been widely applied to empirical studies of dialogical argumentation, especially in domains characterized by rich argumentative exchanges. The paper (Lawrence et al. 2017) demonstrated its utility in analyzing radio broadcasts and mediation dialogues, where turns and argumentative structures could be systematically mapped. Subsequent studies focused on political discourse, taking advantage of the IAT's capacity to model interactional patterns in formal debate. For example, Duthie and Budzynska (2019) analyzed the British Parliament Sessions (Hansard) to uncover conflict and inferential patterns in institutional dialogue. Similarly, in my own work (Konat et al. 2016b) we applied IAT to a US legislative forum, illustrating how arguments are co-constructed across contributions from different participants. With Visser et al. (2020) we extended this approach to analyze both the pre-election debates of 2016 and audience reactions on Reddit, exploring the dialogical dynamics of real-time argumentation.

IAT is a methodology that resonates with my approach: its diagrams, which represent both dialogical and argumentative dimensions of dialogue, are admittedly complex but highly effective in revealing the anchoring of argumentation in dialogical moves. The reader may anticipate where I am heading here: towards the interactional model of pathos I propose in this book. Just as argumentation in both the IAT and the PD framework is inherently interactive, so too is pathos: there is no purpose in performing an emotional appeal without an interlocutor to whom the appeal is directed for persuasive gain. Although IAT itself does not explicitly engage in emotional appeals, the model proposed in this book integrates seamlessly with its methodology. For reasons of clarity and focus, I do not introduce the full convention of IAT diagramming here; however, readers interested in these structures will find them developed in detail in my other works. The conceptual foundation remains the same: IAT conceives of argumentation in the form of a "*p* because *q*" structure, and I adopt this understanding throughout the present volume. For me, pathos is not itself an argument, which, I believe, is a view that can at times be misleading. Rather, pathos accompanies logos or becomes interwoven with it. From the perspective of analysis, and in IAT terms, arguments are reconstructed as structures of propositions derived from locutions. In my approach, emotional appeals may manifest in specific lexical choices or phrasings employed by the speaker to enhance the persuasive force of their argumentative structure, or in other forms surrounding the argument structures. After presenting my model of interactional pathos in the next chapter, the remainder of this book will be dedicated to identifying such emotional appeals within argumentative discourse.

3.4 Introducing Model of Interactional Pathos

3.4.1 The Model of Interactional Pathos in Argumentation (MIPA)

To summarize the discussion so far: first, argumentation is an interactive process that takes place between two or more individuals. This perspective is supported by both Pragma-Dialectics (PD) and Inference Anchoring Theory (IAT). In PD, argumentation is analyzed as occurring, for example, during a critical discussion aimed at resolving differences of opinion. Similarly, IAT holds that argumentation must be "anchored" between two dialogical moves. In other words, while reasoning can be done individually, argumentation arises between people: either when a counterpart expresses doubt (or, in IAT terms, challenges a claim by asking "Why *p*?") or when such a challenge is anticipated. In some cases, the interlocutor may be an audience before whom we present the reasons for our claims. Consequently, in this book, I understand arguments as structures such as "*p* because *q*" that appear in dialogical or other interactive contexts.

Following from this, second, we focus here mainly on persuasive argumentation, i.e., argumentation that occurs not in any type of dialogical interaction but specifically in situations where one interlocutor is attempting to persuade others that their claim is true. Therefore, persuasive argumentation is (as previously illustrated with Tokarz's distinction between argumentation, persuasion, and manipulation) argumentation performed with the aim of changing someone's opinion. By this, it differs from argumentation in scientific inquiry or critical discussion contexts, which aim at unveiling the truth or resolving disagreements.

Third, we observed that the topic of emotions in language is extensively studied from the perspective of linguistic pragmatics and argumentation theory. Pragmatic accounts follow the intuition that emotions constitute either a context for argumentation or are used to manipulate such contexts. Argumentation theorists include emotional appeals, emotional expressions, and emotions as arguments into their work more and more, as they look into cases of natural language argumentation.

Taking all this into account, I can now propose the conceptualization of pathos used in this book:

Definition 3.1 (Pathos) Pathos is an emotional appeal performed in a persuasive interaction by the speaker with the aim of eliciting emotions in the audience in order to increase the perceived strength of the speaker's argumentation.

Let us look closer at the elements of this concept. First, we can observe that pathos covers only emotional appeals, leaving aside all other forms of how emotions can be present in argumentative contexts, cases that have been mentioned in earlier chapters: expressed emotions, emotions as arguments, said emotions, and argued emotions. All of these, while constituting an important strand of emotion research, are separate from pathos and will not be used in this book.

How does emotional appeal work? We can imagine that the speaker, with the use of their words, is "reaching" into the mind of the hearer and attempting to elicit such an emotional state that this state provides a background for the arguments presented later, making these arguments appear to the hearer as stronger. This is regardless of their actual soundness—even though the use of pathos is sometimes associated with weak arguments as a technique aimed at covering this weakness, emotional appeal can also accompany strong arguments. There is no reason for a speaker not to appeal to emotions, even if they have strong arguments at hand. The use of emotional appeal does not automatically cause an argument to be fallacious, as Walton has extensively demonstrated (1992).

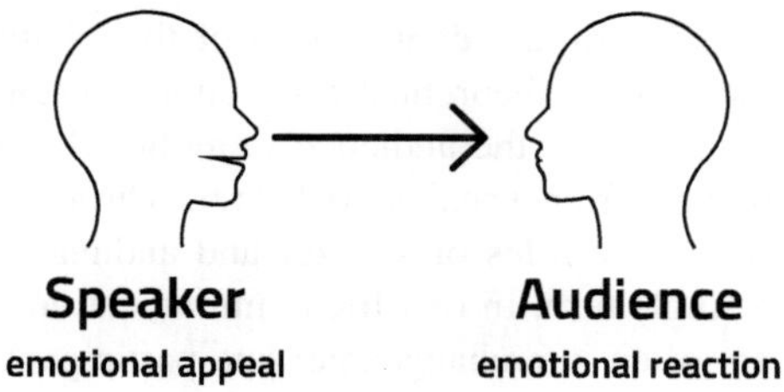

Fig. 3.1 Model of Interactional Pathos in Argumentation (MIPA): general structure of emotional appeal in speaker–audience interaction

The root of my concept of pathos lies in Aristotle's idea of appealing to the emotions of the audience. However, what I propose here is an update that includes modern conceptualizations taken from the study of persuasion, pragmatics, and argument strength. As we shall see throughout this book, this integration allows for further developments in both theoretical and empirical manners. As proof of concept, I will start with my proposed interpretation of this modern pathos concept within speech act theory and later present the first in a series of empirical studies putting this concept to the test in both psychological and computational contexts.

The Fig. 3.1 presents the general form of the interactive model of pathos proposed in this book. It is a developed version of the early sketch of the model introduced in my previous work (Konat et al. 2024). The central idea draws from the Aristotelian view that pathos arises between the speaker (on the left) and the audience or hearer (on the right).

On the cognitive level, the speaker intends to perform a pathos appeal to evoke a specific emotional state in the audience for persuasive purposes. This can be achieved through various linguistic means. The specific methods used by speakers to realize pathos appeals at the language level will be examined in the following chapters, based on experimental data and corpus studies.

On the audience's side, the cognitive experience involves an emotional reaction to the pathos appeal. These reactions may align with the speaker's intention, contradict it, or show no clear emotional response. We will explore the different ways such reactions are manifested, through linguistic expressions of emotion, mood changes, or shifts in perceived argument strength. The arrow from speaker to hearer indicates the interactive nature of pathos. The speaker metaphorically reaches into the hearts and minds of the audience, attempting to influence their emotional states and create conditions more favorable for argument reception.

It is important to treat this model as an idealization. It does not capture all aspects of real-world persuasion, such as the influence of non-verbal cues, ethos appeals by the speaker, or the audience's pre-existing emotional states. Emotional appeals and reactions are not limited to argumentation but may occur in other parts of a dialogue. These limitations acknowledged, the model provides a workable basis for operationalizing pathos in psycholinguistic and computational contexts. This will be explored in a series of studies that employ adapted forms of MIPA in the following parts of the book.

The primary contribution of this chapter is the introduction of the model and the explicit theoretical distinction it establishes between the speaker's appeals to emotion and the audience's emotional reactions. It represents only one dialogical move—from speaker to hearer. Throughout this book, we focus on such cases, where the roles of speaker and audience are fixed and the interaction consists of a single turn. In real-life communication, especially in online contexts, these roles are often interchangeable: one participant may be a rhetor in one exchange and a hearer in another. Modeling these more complex interactions is a goal for future research. Before that, we need a clearer understanding of how individual pathos appeals function in communication.

3.4.2 *Speech Act Interpretation of MIPA*

In terms of Austin (1975), what is the speaker doing by performing an emotional appeal? Let me provide a simple breakdown of the elements of an emotional appeal in terms of speech acts, starting with the concept of illocution:

Definition 3.2 (Illocution of Emotional Appeal) The intention of eliciting an emotional reaction by the speaker in the hearer for persuasive gain.

Therefore, just as in the classic Austin (1975) sense, when we can say what the speaker is doing with words (asking a question, pronouncing a sentence, making an appointment), in the case of pathos, the speaker seeks to elicit emotions. It is analogous to Austin's sense of the illocution of asking a question as seeking to elicit an answer. Persuasive gain in this context is understood as an increase in the argument strength perceived by the hearer.

Definition 3.3 (Locution of Emotional Appeal) The deliberate choice of emotion-eliciting wording by a speaker, where neutral phrasing could have conveyed a similar proposition that could also serve as a reason in the argument being made.

Figure 3.2 presents a more detailed version of the graphic representation of the general pathos model, with added placing of illocution, locution, and perlocution.

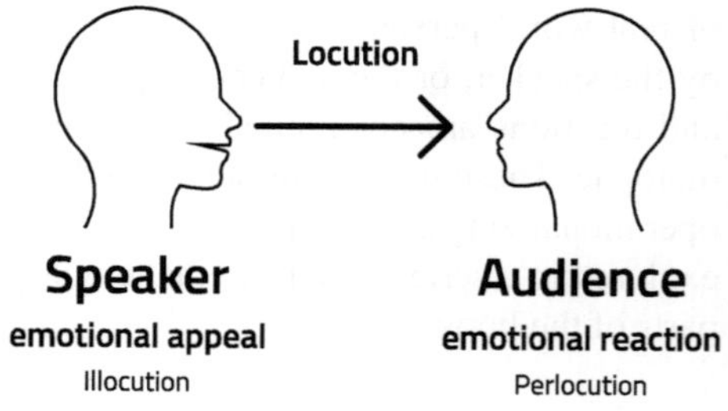

Fig. 3.2 Model of Interactional Pathos in Argumentation (MIPA): structure of emotional appeal as a speech act (illocution, locution, perlocution)

In the context of the communicative act, while emotions are typically associated with the expressive function of language, understood as the manifestation of the internal state of the speaker, emotional appeals in argumentation are primarily impressive in nature. That is, their function is not to express how the speaker feels, but to influence how the audience feels or responds. This distinction is central for modeling pathos as a speech act: the emotional appeal is not an act of self-expression, but a strategic move oriented toward persuasive effect.

As an illustration, let me go through example (1) from the Introduction: "Social media leads to addictions; these are very simple addictions and often not taken as seriously as, for example, alcohol or cigarettes" (*q*). This is an utterance from a student who argues for the thesis "We should quit social media" (*p*) during an experimental setting. The details of this experiment will be provided in Part II; here, let us just observe that it fulfills the conditions of a persuasive context (the student is involved in a competitive debating situation) and at the same time creates an argument structure "*p* because *q*". The speaker deliberately decides to use specific wording, assuming this wording will elicit emotions (probably fear): "addictions", "alcohol", "cigarettes". By comparing social media use to substance abuse, the speaker taps into pre-existing emotions of the hearers, who are aware that those substances can be devastating to one's health, and seeks to elicit those emotions in that moment to increase the perceived strength of their argument.

At the same time, as a justification for the conclusion "We should quit social media", a number of more neutral premises would also be adequate purely from an argumentative point of view. For example: "Social media fosters habits that, while often dismissed as insignificant, can be pervasive and problematic" (*r*). In terms of propositional content, proposition *r* is quite similar to *q*, or at least similar enough to serve the same function within a given argument structure. This observation is captured in the concept of rephrase in argumentation, which was developed by myself and colleagues in earlier works (Konat et al. 2016a; Koszowy et al. 2022) and is still currently studied in empirical and corpus approaches (Younis et al. 2023).

In argumentative settings, a rephrase relationship is said to hold between two non-overlapping constituents of argumentation if both constituents occupy the same position in the argument structure and fulfill the same argumentative function (Konat et al. 2016a). Simply speaking, propositions *q* and *r* are in a rephrase relation when one can be swapped for another in an argument structure, and the argument will still hold.

Furthermore, in more recent work (Koszowy et al. 2022), we conceptualized rephrasing in the pragmatic context and partially explored its links to pathos. We observed how speakers often reformulate their claims—seemingly repeating the same propositional content—but in fact adding or removing emotional wording. In the context of the model of pathos in pragmatic settings presented here, rephrase allows me to clearly present the role of locution: the speaker's lexical and structural choices not only convey propositional content but are purposefully selected to elicit emotions in the hearer. This leads us to the last element of Austin's idea:

Definition 3.4 (Perlocution of Emotional Appeal) The emotion elicited in the hearer, which increases persuasive gain.

In other words, the perlocutionary effect of emotional appeal is not only the emotion felt by the hearer but also the influence of that emotion on the perceived strength of an argument. In the classic Austin sense, emotional appeals can be felicitous (the speaker manages to elicit the emotion they intended, and the presence of this emotion increases the perceived strength of an argument) or not (the speaker does not elicit any emotion, elicits an emotion other than intended, or after successful elicitation of emotion, it causes persuasive loss rather than persuasive gain).

This formulation of pathos in terms of speech act theory in the classic Austin sense allows for operationalization in the studies presented throughout this book. As researchers using corpus techniques, i.e., looking at a body of text produced by speakers, what we actually have access to are locutions. With the use of manual annotation, we attempt to reconstruct illocutions—the original intentions of the speaker—by asking, "In this utterance, does the speaker intend to elicit emotions for persuasive gain?" Studies based on this type of annotation will be presented in the next chapter, and in Part III, I will propose a development of this method into automated pathos mining. As we will see, LLMs (often called AI), despite having no intentions or emotions, are able to effectively emulate the behavior of human annotators in decoding illocutions based on locutions present in corpora.

To assess whether an emotional appeal is effective, i.e., to capture the perlocutionary effect, we need to move away from studying just the speaker's side and move towards analyzing the audience's reactions. To this end, I will present two methodologies. In the next chapters, we will observe how audience reactions can be captured via social media, and based on the assumption that people express their emotional reactions accurately, we can assess how effective the emotional appeals of politicians are. To fully control the felicitousness of emotional appeals as speech acts, my team also conducted a series of psychological experiments where we captured audience reactions with questionnaires. These experiments will be described in Part II of this book.

To sum up, the modern reformulation of the concept of pathos—MIPA—that I propose here is rooted in Aristotle but also pragmatic theories, both original and those already incorporated into argumentation theory. Pathos is an interactional process in which the speaker "reaches" with their words to the mind of the hearer with the aim of persuasive gain.

While this concept of pathos is highly idealized, omitting factors such as non-verbal communication, pre-existing emotional states, and emotional expressions outside of argumentation, it offers significant advantages. By focusing on the linguistic and cognitive elements of pathos, this model allows for the operationalization of emotional appeals, making systematic, both computational-level and experimental analysis of emotions within argumentation possible, as will be presented in Parts II and III of this book.

3.5 Model of Interactional Pathos in Argumentation in Pre-election Debates Annotation

Building on the theoretical model presented in previous section (MIPA) we can now present one of the application of such model to observation of patterns of pathos in natural language argumentation. In this section I present manual annotation of instances of argumentation scheme in the sample of transcripts of pre-election debates from Poland and US, and results of this analysis, as previously presented in my team's paper (Konat et al. 2024).

We adopt Walton, Reed, and Macagno's taxonomy of Argument Schemes (2008), customizing it to identify instances of pathos-related Argument Schemes in political debates. These include Arguments from Positive and Negative Consequences, Fear and Danger Appeal, Threat, Slippery Slope, Waste, and Need for Help. The selection of the eight Argument Schemes as pathos-related is our own categorization, not that of Walton, Reed, and Macagno.

The diagram in Fig. 3.3 is the first in a series presenting applications of the pathos model used throughout this book. At this stage, the focus is on the speaker's side. By analyzing the linguistic surface level, we aim to uncover the intention behind emotional appeals through the identification of argument schemes. To improve clarity, the structure is divided into two levels of description for each participant, speaker and audience. The first, labeled Cognitive level, contains 'emotional appeal' for the speaker and 'emotional reaction' for the audience. The second, labeled Language level, presents operationalizations of these concepts as they appear in the various studies discussed. In this first case study, the speaker is represented by a male and female figure in formal clothing, indicating political speakers. On the Language level, this study focuses on argument schemes, which serve as the linguistic realization of the emotional appeal. This study is limited to the perspective of the speaker.

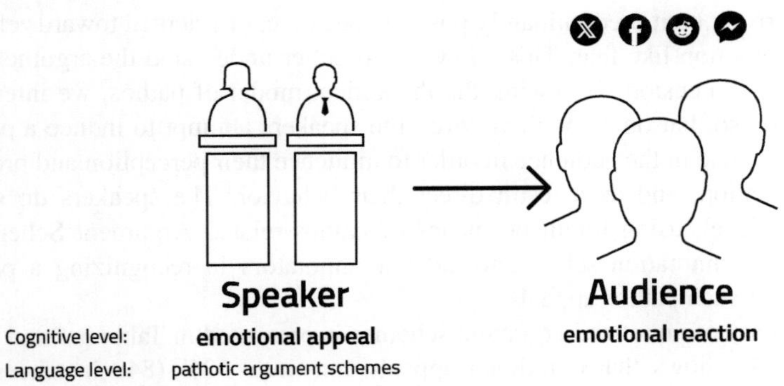

Fig. 3.3 Model of Interactional Pathos in Argumentation (MIPA): integration with pathotic argument schemes. Schematic showing how MIPA integrates with a taxonomy of pathotic argument schemes in persuasive contexts

Table 3.1 Results of the Argument Schemes manual annotation study

Argument scheme	Number	Percentage
Non Pathotic Arguments	772	47.6
Positive Consequences	340	21.0
Negative Consequences	299	18.4
Need For Help	81	5.0
Danger Appeal	69	4.3
Fear Appeal	44	2.7
Causal Slippery Slope	12	0.7
Waste	4	0.2

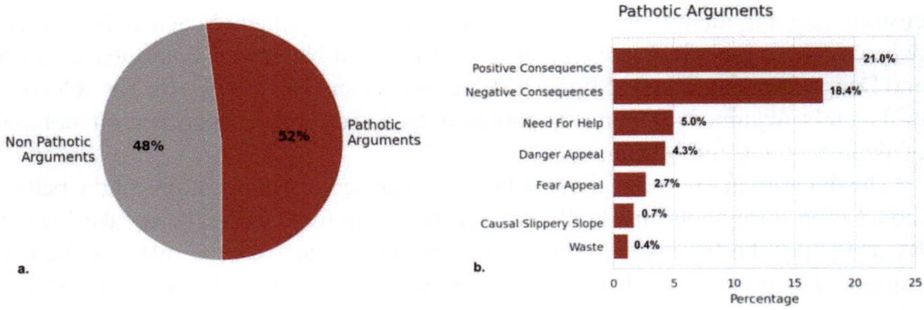

Fig. 3.4 Pathotic schemes distribution in Argument Schemes manual annotation study

The annotation scheme proposed in the current work is based on the critical work on argument schemes (Walton et al. 2008). However, in our operationalization we have revised the argumentation schemes to concentrate on the central emotional intent ("the message of the argument" in our annotation scheme). By this, we mean the primary emotional tone or effect that the speaker aims to evoke, which is integral for this argument. Upon identifying whether an emotional intent is central to the argument, we further categorize this into its "essence", i.e. dominant emotional characteristic, be it predominantly positive, negative, or oriented towards eliciting a specific emotion like fear. This allows us to better understand the argument's core emotional dimension. Following the theoretical model of pathos, we interpret the interaction so that on a cognitive level, the speakers attempt to induce a particular emotional state in the audience in order to influence their perception and processing of information, and as a result direct their behavior. The speakers do so, on a language level, using linguistic means of pathos-related Argument Schemes. We design our annotation scheme to aid our annotators in recognizing a particular instance of such pathos appeals.

Summary of annotated argument schemes is presented in Table 3.1 and Fig. 3.4. Figure 3.4a shows that emotional appeals constitute 52% (849 instances) of all the arguments in the analysed debates, whereas non-pathotic arguments represent 48.0% (772) of the arguments in the final corpora, which means that they were considered by annotators as not appealing to pathos. The summary presented in

Fig. 3.4b indicates that the two most frequent arguments related to pathos are arguments from Positive Consequences (340 arguments; 21.0%) and arguments from Negative Consequences (299; 18.4%). Arguments from Need for Help, Argument from Danger Appeal, and Argument from Fear Appeal constitute respectively 5.0% (81), 4.3% (69), and 2.7% (44) of the re-annotated corpora. Slippery Slope arguments are present in 0.7% (12) of the sample and Argument from Waste was assigned to 0.2% (4) of the arguments. Argument from Threat was not annotated by the raters in any case.

The inter-annotator agreement (IAA) was fair to moderate across different Argument Schemes, with the highest agreement observed in Arguments from Negative Consequences. This suggests a reasonable level of consistency in the identification of emotional appeals in the debates. The analysis also indicated a 'long tail' distribution in the frequency of different Argument Schemes, with general categories like Positive/Negative Consequences being more prevalent.

This study demonstrates the applicability of the MIPA model in identifying emotional appeals in naturalistic political discourse with manual annotation of argument schemes. By customizing Walton et al.'s taxonomy to pathos-related schemes, we provide empirical evidence that emotional argumentation constitutes a significant portion of pre-election debates. The high frequency of arguments from Positive and Negative Consequences confirms the central role of evaluative and affective reasoning in persuasive political communication.

This chapter has introduced the Model of Interactional Pathos in Argumentation and demonstrated its practical application. By establishing a clear distinction between the perspectives of speakers and hearers, we have been able to operationalize pathos appeals on the speaker side as observable phenomena at the language level. This operationalization, I believe, accurately reflects the underlying communicative intentions at the cognitive level. However, it is important to be cautious in interpreting these results due to the moderate levels of agreement between annotators, a reminder of the inherently subjective nature of this phenomenon.

As we conclude Part I of this book, the richness and complexity of emotion in reasoning become evident. Building upon established theories and introducing a new model of pathos, I looked into the exploration of pathos within the realms of ancient and modern rhetoric, argumentation theory and pragmatics. As these disciplines were moving closer to everyday argumentation, scholars representing them were becoming more and more interested in the role of emotional content. Notwithstanding certain critics, emotions now have an acknowledged role in the field. Building on this, I propose a new model of pathos, MIPA, and a series of its operationalizations, illustrating its versatility in different methodologies. This model aims to provide a clearer understanding of the complex relationship between speaker and audience in the context of emotional appeals. As we close this section, we move to the next part of the book, which will look closer into the studies of language of emotions, as well as the definitions and findings from the discipline most closely associated with emotions: psychology.

References

Austin JL (1975) How to do things with words. Oxford University Press

Budzynska K, Reed C (2011a) Speech acts of argumentation: Inference anchors and peripheral cues in dialogue. In: Workshops at the Twenty-Fifth AAAI Conference on Artificial Intelligence

Budzynska K, Reed C (2011b) Whence inference. University of Dundee Technical Report

Duthie R, Budzynska K (2019) Classifying types of ethos support and attack. In: Computational Models of Argument, Frontiers in Artificial Intelligence and Applications, vol 305, IOS Press, pp 161–168, https://doi.org/10.3233/978-1-61499-906-5-161

van Eemeren FH, Grootendorst R (1984) Speech Acts in Argumentative Discussions: A Theoretical Model for the Analysis of Discussions Directed towards Solving Conflicts of Opinion, Studies in Pragmatics, vol 1. Foris Publications, Dordrecht

van Eemeren FH, Garssen B, Krabbe ECW, Henkemans AFS, Verheij B, Wagemans JHM (2014) Handbook of Argumentation Theory. Springer, Dordrecht, https://link.springer.com/referencework/10.1007/978-90-481-9473-7

Gilbert MA (2005) Let's talk: Emotion and the pragma-dialectic model. In: van Eemeren FH, Houtlosser P (eds) Argumentation in Practice, John Benjamins, Holland, pp 43–52

Grice HP (1975) Logic and conversation. In: Cole P, Morgan JL (eds) Syntax and Semantics, Volume 3: Speech Acts, Academic Press, pp 41–58

Hautli-Janisz A, Budzynska K, Reed C (2023) Conventional implicatures in argumentation. Languages 8(1):14, https://doi.org/10.3390/languages8010014, https://www.mdpi.com/2226-471X/8/1/14, reprinted in Oswald (ed.), *Pragmatics and Argumentation*, 2024, pp. 79–104.

Hinton M (2021) Evaluating the Language of Argument, 1st edn. Argumentation Library, Springer Cham, https://doi.org/10.1007/978-3-030-61694-6, published: 01 December 2020 (hardcover), 30 November 2020 (eBook)

Huang Y (2020) Pragmatics. https://plato.stanford.edu/entries/pragmatics/#Thepr, accessed: 23.20.23

Konat B, Budzynska K, Saint-Dizier P (2016a) Rephrase in argument structure. In: Saint-Dizier P, Stede M (eds) Foundations of the Language of Argumentation. COMMA 2016 Workshop, University of Potsdam, pp 32–39

Konat B, Lawrence J, Park J, Budzynska K, Reed C (2016b) A Corpus of Argument Networks: Using graph properties to analyse divisive issues. In: Proceedings of the 10th edition of the Language Resources and Evaluation Conference (LREC 2016)

Konat B, Gajewska E, Rossa W (2024) Pathos in natural language argumentation: Emotional appeals and reactions. Argumentation 38:369–403, https://doi.org/10.1007/s10503-024-09631-2

Koszowy M, Oswald S, Budzynska K, Konat B, Gygax P (2022) A pragmatic account of rephrase in argumentation: linguistic and cognitive evidence. Informal Logic 42(1):49–82, https://doi.org/10.22329/il.v42i1.7212

Lawrence J, Visser J, Reed C (2017) Harnessing rhetorical figures for argument mining. Argument & Computation 8(3):289–310

de Oliveira Fernandes D, Oswald S (2023) On the rhetorical effectiveness of implicit meaning–a pragmatic approach. Languages 8(1):6, https://doi.org/10.3390/languages8010006, https://www.mdpi.com/2226-471X/8/1/6, reprinted in Oswald (ed.), *Pragmatics and Argumentation*, 2024, pp. 23–78.

Oswald S (2010) Pragmatics of uncooperative and manipulative communication. PhD thesis, Université de Neuchâtel, Institut des Sciences du Langage et de la Communication

Oswald S (2023) The pragmatics and argumentation interface. Languages 8(3):210, https://doi.org/10.3390/languages8030210, https://www.mdpi.com/2226-471X/8/3/210, reprinted in Oswald (ed.), *Pragmatics and Argumentation*, 2024, pp. 1–22.

Reed C, Walton DN, Macagno F (2007) Argument diagramming in logic, law and artificial intelligence. The Knowledge Engineering Review 22(1):87–109, https://doi.org/10.1017/S0269888907001051

Reijven MH (2024) “so i know how to do this”: The prototypical argumentative pattern in u.s.a. presidential debates. Journal of Language Aggression and Conflict 12(1):89–110, special issue: Pragmatic perspectives on disagreement

Searle JR (1979) Expression and meaning: Studies in the theory of speech acts. Cambridge University Press

Sperber D, Wilson D (1996) Relevance: Communication and Cognition, 2nd edn. Blackwell

Tindale CW (2022) Utterer meaning, misunderstanding, and cultural knowledge. Languages 7(3):172, https://doi.org/10.3390/languages7030172, https://www.mdpi.com/2226-471X/7/3/172, reprinted in Oswald (ed.), *Pragmatics and Argumentation*, 2024, pp. 285–305.

Toulmin S (1958) The Uses of Argument. Cambridge University Press, Cambridge

Visser J, Konat B, Duthie R, Koszowy M, Budzynska K, Reed C (2020) Argumentation in the 2016 US presidential elections: annotated corpora of television debates and social media reaction. Language Resources and Evaluation 54(1):123–154

Wagemans JHM (2016) Constructing a periodic table of arguments. In: Bondy P, Benacquista L (eds) Argumentation, Objectivity, and Bias: Proceedings of the 11th International Conference of the Ontario Society for the Study of Argumentation (OSSA), 18–21 May 2016, OSSA, Windsor, ON, pp 1–12, retrieved from https://scholar.uwindsor.ca/ossaarchive/OSSA11/papersandcommentaries/190/

Wagemans JHM (2023) How to identify an argument type? on the hermeneutics of argumentative discourse. Journal of Pragmatics 203:117–129, https://doi.org/10.1016/j.pragma.2022.12.011

Walton D (1992) The Place of Emotion in Argument. The Pennsylvania State University Press, University Park, PA

Walton D, Reed C, Macagno F (2008) Argumentation Schemes. Cambridge University Press, Cambridge, https://doi.org/10.1017/CBO9780511802034

Wharton T, de Saussure L (2023) Pragmatics and Emotion. Cambridge University Press

Wharton T, Bonard C, Dukes D, Sander D, Oswald S (2021) Relevance and emotion. Journal of Pragmatics 181:259–269

Younis R, Fernandes DdO, Gygax P, Koszowy M, Oswald S (2023) Rephrasing is not arguing, but it is still persuasive: An experimental approach to perlocutionary effects of rephrase. Journal of Pragmatics 210:12–23, https://doi.org/10.1016/j.pragma.2023.03.010

Part II
Language and Psychology of Pathos

Chapter 4
The Language of Emotion

Abstract This chapter explores how emotions are expressed, named, and strategically deployed in argumentative discourse. Drawing on discourse analysis, rhetorical theory, and Pragma-Dialectics, it examines the functions of emotion terms, the role of emotional appeals in dialogical action, and the interaction between metaphor and affect. The chapter begins by tracing the French tradition of analyzing emotion naming as a marker of speaker involvement, commitment, and evaluative stance. Then it turns to dialogical perspectives, showing how emotions function as both argumentative resources and potential sources of fallacy. A review of rhetorical and computational approaches illustrates how emotions are increasingly recognized as integral to argument quality, audience adaptation, and persuasion. Finally, the chapter analyzes the role of metaphor as a conceptual and affective device in argumentation. An empirical study of political discourse demonstrates that metaphor and emotional appeal frequently co-occur, suggesting that speakers select emotionally resonant source domains to strengthen the persuasive force of arguments.

Keywords Emotion terms · Emotional appeals · Argumentation · Rhetorical strategies · Metaphor and affect · Political discourse

4.1 Naming Emotions in Argumentation

Emotion naming plays a crucial role in argumentative discourse, particularly in its capacity to signal the speaker's stance and to guide the audience's evaluative orientation. In the French tradition of discourse analysis, several authors have examined the functions and rhetorical roles of emotion terms in argumentative structures.

In his foundational work, Plantin (1999) introduces the idea of "emotion terms" into argumentation. In Plantin (2019) he emphasizes tension as a discourse-level phenomenon reflecting speaker involvement. Unlike traditional views that associate tension with negativity, he conceptualizes it as a sign of the speaker's commitment, aimed at inducing shared involvement from the audience. Tension thus operates as a

B. Konat, *Emotional Appeals in Argumentation*, Argumentation Library 47,
https://doi.org/10.1007/978-3-032-06332-8_4

persuasive device, closely linked with emotional arousal and rhetorical intensity. Again, we can see here what was already present in previous approaches: that emotions do not have to impede the dialogue, but are a natural part of speaker's involvement. Moreover, Plantin identifies linguistic markers of tension, including exclamations, rhetorical questions, and radicalization of argument. These markers heighten emotional salience and often serve as a basis for attributing emotions such as anger or shame. The presence of "emotion terms" in exclamatory outbursts is treated as a linguistic indicator of attributed emotions.

Plantin's approach later informed multiple applications in French tradition and worldwide. For example, models of group emotions in deliberation are proposed in Polo et al. (2017). These models conceptualize emotions as cognitive and social resources within argumentation, distinguishing between felt emotions and emotional tonalities attributed to discourse objects. This distinction aligns with the analytical framework developed in this volume.

In works of Sara Greco's team, naming emotions plays a central role in conflict resolution and deliberative practices. The main focus is on what they call "euphoric" and "dysphoric" emotions. In the paper Greco et al. (2024) mediators in dispute contexts identify and explicitly name disputants' emotions, transforming them into "said" emotions. This discursive move clarifies the conflict's core, positions the parties as co-arguers, and brings emotions into argumentative discussion. Mediators begin by identifying dysphoric emotions at the root of the conflict, making them explicit, and then guide the discussion along a trajectory from dysphoric to euphoric states. This approach informs the analysis presented in Greco and Jermini-Martinez Soria (2024), which focuses on online deliberative platforms. Here, the presence of dysphoric emotions signals a need for conflict prevention. The authors argue that complex political topics discussed online often evoke strong emotional reactions, as participants perceive something personally important to be at stake, making such discussions prone to conflict. They propose conflict prevention tools centered on human moderation. Moderators are advised to ask open and non-accusatory questions that invite participants to explain the reasons behind their feelings. In addition, when dysphoric emotions involve guilt or resentment, moderators may reframe the issue from an individual concern to a shared or systemic one. These interventions, derived from dispute mediation, are adapted to support emotionally sensitive engagement in online deliberative spaces. Again here, we observe a focus on the language of emotions, but also on the versatility of emotion models applied across face-to-face and online interactions. This is the trajectory followed in this book, which demonstrates how the proposed Model of Interactional Pathos in Argumentation (MIPA) functions in direct, online, and mixed dialogical contexts across several languages.

In another study continuing the French tradition, Cigada (2019) focuses on the discursive function of the emotion terms "appreciation" and "tension" in the discourse following the Charlie Hebdo attack. These are examined as integral components of argumentative conclusions. For instance, the phrase "had expressed appreciation" (Fr. "s'était félicité") is used to convey a positive emotional evaluation by students. Drawing on Plantin's classification of emotion terms, Cigada reconstructs this expression as an indicator of joy or satisfaction.

Micheli (2010) and Herman and Serafis (2019) provide a broader typology of emotion in discourse (2010, 2019). Micheli contrasts two perspectives on emotions in argumentation: the standard view treats them as adjuncts to persuasion, while the alternative view regards emotions as the core object of deliberation. In the latter, speakers argue about the legitimacy of specific emotions, thereby making emotions themselves the claim under debate. This shift positions emotion naming not as persuasive embellishment but as a substantive component of argumentative reasoning.They differentiate 'said", "shown," and "argued" emotions. "Said" emotions are lexicalized, "shown" emotions are inferred from paralinguistic or syntactic cues, and "argued" emotions are constructed through discursive representations of emotionally charged situations. These typologies clarify how emotion is encoded and inferred in argumentative texts, a view that is present also in other works. For example Mazzali-Lurati et al. (2019) and van Haaften (2019) demonstrate how hyperbole and emotive intensifiers function as strategic tools to attract attention and enhance argument salience. The presence of multiple emotive terms can exaggerate evaluative stances, influencing hearer perception and evaluative inference.

In sum, naming emotions in argumentation serves not only to express or elicit affective responses but also to structure the interaction and shape evaluative perspectives. The French tradition and related approaches illustrate how emotion terms, whether lexicalized, inferred, or discursively constructed, anchor argumentative positions, manage conflict, and signal commitment.

4.2 Emotions in Dialogical Action

This section synthesizes multiple theoretical perspectives on the role of emotions in dialogical action, especially within the context of argumentation. The contributions span informal logic, rhetorical theory, Pragma-Dialectics, and computational approaches, converging on the view that emotions are not always disruptive but can play a substantive, even constructive, role in argumentative discourse.

One major line of inquiry addresses the normative and strategic status of emotional appeals in argumentation, a status that remains contested across traditions. As Manolescu (2006) explains, while some informal logicians view emotions as irrelevant to rational argumentation, others advocate the legitimacy of "pathotic arguments" provided that they are grounded in reasonable cognitions. According to her, Pragma-Dialecticians typically classify emotional appeals as fallacies due to a failure of relevance. However, Manolescu points out that PD, particularly through its theory of strategic maneuvering, allows for the analytical integration of emotional appeals when they align with audience adaptation and stylistic considerations. This flexibility allows for the evaluation of emotional appeals as potentially relevant under specific dialogical conditions. This continuation of Walton's position (outlined in Part I) finds resonance in Manolescu's work, where she applies the theoretical concepts of emotional argumentation within Pragma-Dialectics to language analysis. Her analysis of Frederick Douglass' speech illustrates how emotional appeals such as shame and indignation can be reconstructed as logically good arguments when grounded in acceptable and contextually appropriate premises. The appeal aligns with situational norms, demonstrating that emotional intensity can function rhetorically to support premise adequacy.

Further analysis focuses on emotional appeals as fallible but strategically viable elements of dialogue. In her subsequent work, Manolescu (2007) looks into the specific case of shame appeals. Drawing on Walton's framework, she argues that such appeals are not inherently fallacious but must be evaluated according to their inferential structure and dialogical context. When properly situated, shame may function as a prudential reason for action. However, excessively intense or contextually inappropriate appeals risk disrupting dialogue and undermining the arguer's credibility.

In a related strand, other studies emphasize the regulation of emotional dynamics in deliberative interactions. González González et al. (2019) bring attention to the dynamics of emotion regulation in argumentative dialogue. They observe that emotional appeals, whether directed at the speaker or the listener, affect memory, attention, and situation assessment, all of which influence deliberation. They draw on theories by Meyer and Tindale to argue that emotions can reflect values and are thus susceptible to argumentative modulation. Effective argumentation requires not only rhetorical skill but also emotional self-regulation, especially in interactive or deliberative contexts.

Shifting the focus to the form of argumentation itself, visual rhetoric examines a different medium through which emotional appeals operate. Kjeldsen (2015) explores rhetorical argumentation in visual media, arguing that emotionally charged images can evoke immediate responses that bypass deliberation. Yet he is cautions against equating emotional imagery with irrationality. Like verbal appeals, visual arguments must be evaluated for their relevance, sufficiency, and acceptability in context. Walton (2010) similarly examines visual arguments in advertising, noting that emotional content can legitimately shift burdens of proof in practical reasoning, provided the appeals are contextually justified and do not distort the speaker's position.

Turning to recent developments, the computational turn in argumentation studies has begun to address emotional elements, though not comprehensively. As we will see in detail in Part III of this book, contemporary computational models of argumentation have begun to acknowledge the importance of emotional appeals, though research remains sparse. Lawrence et al. (2017) and Green (2020) explore how rhetorical figures and emotional cues can enhance argument mining and computational persuasion. Hunter (2018) differentiates normative argumentation from persuasive effectiveness, stressing that emotional appeals can be powerful, particularly when tailored to audience traits. Still, current systems overemphasize logical coherence at the expense of the emotional component. Integrating emotional modeling into computational persuasion remains a challenge for the field.

Relatedly, computational models of dialogue structure suggest further integration of emotion as part of argumentation acts. Pallotta and Delmonte (2011) propose a dialogue-oriented approach to argumentation, recognizing communicative acts such as acceptance, rejection, and justification within dialogical structure. Although not focused on emotion per se, their schema accommodates emotionally inflected moves (e.g., shaming, disapproval) that align with observed rhetorical patterns. Their work suggests the potential for integrating affective dimensions into dialogical models of argument.

In conclusion, across rhetorical, Pragma-Dialectical, and computational paradigms, there is a growing recognition that emotions are not just irrational forces to be excluded from argumentation but are integral to how arguments function in real-world discourse. However, the literature suggests, that their legitimacy depends on contextual appropriateness, inferential support, and dialogical coherence. Thus, a better understanding of emotional argumentation requires interdisciplinary integration across logic, rhetoric, pragmatics, and computational modeling of language of emotion. This is a conceptual groundwork for the next section, which turns to the role of metaphor. After that, in the following chapter, we will further develop the focus on language of emotion by examining psychological models and their operationalization in affective lexicons.

4.3 Metaphor and Emotions

One linguistic and rhetorical device that closely intersects with emotional appeal is metaphor, a figure deeply connected to affect and conceptual framing. Below, I present selected points made by other scholars that inspired our own study outlined at the end of this section.

According to Ervas (2019), metaphor constitutes a cognitive process through which an unfamiliar conceptual domain (the target) is understood in terms of a more familiar one (the source). This process is not confined to verbal language but extends to visual and multimodal forms, reinforcing the claim that metaphor is fundamentally conceptual. A key element in metaphor comprehension is ignorance: interpreters may either disregard expected features of the source or attribute new, emergent properties, thus generating genuinely novel knowledge. Metaphors act as framing tools that selectively highlight certain aspects of a domain while hiding others, guiding interpretation in a non-neutral direction. Their persuasive function derives from this capacity to reframe a situation, altering conceptualizations and producing new interpretive possibilities. In visual metaphors, the emotional impact is often tied to a "sentiment of (ir)rationality," an epistemic feeling associated with disrupted expectations and their resolution. This process produces a cognitive-emotional reaction, often described as "pleasure or relief", that reflects both conceptual fluency and affective engagement.

Macagno (2020) also approaches metaphor not as a stylistic embellishment, but as an inherently argumentative tool. He argues that metaphors convey arguments or their components and derive persuasive power from their argumentative function. His distinction between arguments "communicated metaphorically" and those "triggered implicitly" through connotation points to the dual function of the metaphor: it can frame arguments explicitly or evoke implicit evaluations. Metaphors direct attention toward salient features of a topic while concealing others, generating "emergent meanings". Connotation carries emotional significance, rooted in commonplaces and cultural associations that provoke indirect emotional responses such as fear. These weak implicatures can extend the scope of evaluative judgment beyond the literal content of the argument.

Santibáñez (2010), offers a framework integrating conceptual metaphor theory with Toulmin's model of argumentation. He treats metaphor as a primarily mental operation that precedes linguistic expression, emphasizing domain mapping over pre-existing similarity. His six-step procedure (from identifying linguistic expressions to establishing domain mappings) enables a systematic analysis of metaphors as elements of argumentative structure. He proposes that conceptual metaphors can serve as backing in Toulmin's model, while proverbial expressions grounded in metaphor may function as warrants. Drawing from Chilean parliamentary discourse, he illustrates how metaphors such as "presidents are mothers/fathers" serve persuasive purposes, albeit acknowledging methodological challenges regarding the identification of conceptual domains.

With slightly different perspective, coming from empirical field, Sopory and Dillard's (2002) provide meta-analytic evidence that metaphorical messages are generally more persuasive than literal ones. The effects noted are small, but recognizable, particularly under conditions that favor novelty, simplicity, and early positioning within a message. The most strongly supported explanation is the superior organization theory, which holds that metaphor enhances coherence and salience in argumentation. While other mechanisms (such as affective responses linked to semantic resolution and perceived communicator credibility) received less empirical support, the emotional dimension remains significant. Metaphors can reinforce message interpretation through affective engagement and are associated with dynamic, persuasive delivery styles.

Applying those perspectives, my team has investigated how emotionally appealing metaphors are deployed in political discourse to elicit affective responses in audiences (Juszczyk et al. 2025). Adopting an interdisciplinary methodology that combines cognitive linguistics, critical discourse analysis, and argumentation theory, we examined metaphor use and emotional appeals in Polish pre-election debates.

(1) WKK: Polish health service is in a *heart–attack threatening condition*, (. . .) it already had two **heart attacks** and will simply not survive the third one. The Polish People's Party proposed the pact for the Polish **health**, for **health**, for the patient.[1]

[1] Pol: Polska służba zdrowia jest *w stanie przedzawałowym*, (. . .) ona jest już *po dwóch zawałach i trzeciego po prostu nie przeżyje*. Polskie Stronnictwo Ludowe zaproponowało pakt *na rzecz polskiego **zdrowia**, na rzecz **zdrowia**, na rzecz pacjenta*. (WKK, TVP, 01.10.2019)

In example (1) from Juszczyk et al. (2025) the speaker employs the argument scheme, in which the premise "Polish health service is in a heart-attack threatening condition..." leads to the conclusion that there is a need for the "pact for the Polish health". He uses conceptual metaphor POLISH HEALTH SERVICE IS A PATIENT, which contributes to the rhetorical cohesion, and appeals to the emotions of the hearers through the word "health" (marked in bold) as eliciting happiness in Polish, following the classification in Wierzba et al. (2021). If we were analyzing this example based on our intuitions rather than on automatic identification of emotional appeals, then we would also identify "heart attack" and "won't survive" as expressions appealing to negative emotions, most likely fear.

Example (1) illustrates persuasive strategies used by politicians. In this type of discourse, the speakers will combine metaphorical expressions and emotion-eliciting words into their argumentation with the aim of greater rhetorical effect, as every utterance in pre-election debates can be considered as having persuasive intent. Based on the literature presented above, we may assume high saturation of metaphors and emotional appeals in political discourse. Hence, the aim of the study from Juszczyk et al. is to analyze the frequency and interaction between three discursive phenomena present in pre-election debates: metaphors, arguments, and emotional appeals, using both concepts and methods from three theoretical traditions and combining them into one cohesive methodology.

In a corpus of 615 arguments from political debates, we found that the majority (49%) combined argumentative content with emotional appeals while excluding metaphors. This was followed by 32% of arguments that included metaphors alongside both argumentative content and emotional appeals. Arguments relying solely on inference without metaphors or emotional appeals accounted for 16% of the data. The least frequent type involved arguments that combined metaphors with inference but did not include emotional appeals, comprising only 3% of the total. These figures suggest that emotional appeals are more frequently combined with arguments than metaphors, either alone or in combination.

What stands out in our findings is that the majority of metaphorical arguments (197 out of 214 cases) also include emotional appeals. This suggests that metaphors on their own occur less frequently in political argumentation. When speakers introduce metaphors, they often accompany them with emotion-eliciting language. Arguments that use metaphors without such emotional elements are rare in our data. This pattern may indicate that, when speakers use metaphors to express abstract concepts, they also make deliberate choices about the source domain to include emotional appeal. In the context of pre-election debates, where persuasion is central, speakers appear to avoid emotionally neutral source domains. Instead, they use emotion-eliciting words to appeal to the minds of the audience. As shown in example (1) the phrase "Polish health service is in a heart threatening condition" is not only metaphorical but also created in a way that might elicit emotions in the audience. The speaker's selection of the source domain functions as a strategic lexical choice aimed at persuasion.

In conclusion, existing literature points to the dual function of metaphor as both a conceptual and affective device in persuasive discourse. Metaphors frame arguments, guide interpretation, and elicit affective reactions, often through carefully chosen source domains that resonate emotionally with audiences. Our empirical findings also support this view, demonstrating that in political communication, metaphor and emotional appeal frequently co-occur.

References

Cigada S (2019) Emotions in argumentative narration: The case of the Charlie Hebdo attack. Informal Logic 39(4):401–431

Ervas F (2019) Metaphor, ignorance and the sentiment of (ir)rationality. Synthese https://doi.org/10.1007/s11229-019-02489-y

González González M, Gómez J, Lemos M (2019) Theoretical considerations for the articulation of emotion and argumentation in the arguer: A proposal for emotion regulation in deliberation. Argumentation 33(3):349–364

Greco S, Jermini-Martinez Soria C (2024) Integrating conflict prevention tools into deliberative democracy online platforms. In: Proceedings of the First Workshop on Language-driven Deliberation Technology (DELITE2024), ELRA Language Resource Association, Torino, Italy, pp 39–44, cC BY-NC 4.0

Greco S, Cigada S, Jermini-Martinez Soria C (2024) The naming of emotions in dispute mediators' strategic manoeuvring: a case study using a french language corpus. Text & Talk 44(1):25–46, https://doi.org/10.1515/text-2021-0044

Green NL (2020) Recognizing rhetoric in science policy arguments. Argument & Computation 11(3):257–268

Herman T, Serafis D (2019) Emotions, argumentation and argumentativity: Insights from an analysis of newspapers headlines in the context of the Greek crisis. Informal Logic 39(4):373–400

Hunter A (2018) Towards a framework for computational persuasion with applications in behaviour change. Argument & Computation 9(1):15–40

Juszczyk K, Konat B, Fabiszak M (2025) The heart attack of the polish health service: metaphors, arguments, and emotional appeals in political debates. Language and Cognition 17:e12, https://doi.org/10.1017/langcog.2024.35

Kjeldsen JE (2015) The rhetoric of thick representation: How pictures render the importance and strength of an argument salient. Argumentation 29(2):197–215

Lawrence J, Visser J, Reed C (2017) Harnessing rhetorical figures for argument mining. Argument & Computation 8(3):289–310

Macagno F (2020) How can metaphors communicate arguments? Intercultural Pragmatics 17(3):335–363, https://doi.org/10.1515/ip-2020-3004

Manolescu BI (2006) A normative pragmatic perspective on appealing to emotions in argumentation. Argumentation 20(3):327–343

Manolescu BI (2007) Shaming in and into argumentation. Argumentation 21(4):379–395

Mazzali-Lurati S, Pollaroli C, Marcantonio D (2019) The rhetorical and argumentative relevance of" extreme consequence" in advertising. Informal Logic 39(4):497–530

Micheli R (2010) Emotions as objects of argumentative constructions. Argumentation 24(1):1–17

Pallotta V, Delmonte R (2011) Automatic argumentative analysis for interaction mining. Argument & Computation 2(2–3):77–106

Plantin C (1999) Arguing emotions. In: Proceedings of the Fourth Conference of the International Society for the Study of Argumentation, Amsterdam, NL

Plantin C (2019) Tense arguments: questions, exclamations, emotions. Informal Logic 39(4):347–371

Polo C, Plantin C, Lund K, Niccolai G (2017) Group emotions in collective reasoning: A model. Argumentation 31(2):301–329, https://doi.org/10.1007/s10503-016-9407-5

Santibáñez C (2010) Metaphors and argumentation: The case of Chilean parliamentarian media participation. Journal of Pragmatics 42(4):973–989, https://doi.org/10.1016/j.pragma.2009.08.019

Sopory P, Dillard JP (2002) The Persuasive Effects of Metaphor: A Meta-Analysis. Human Communication Research 28(3):382–419, https://doi.org/10.1111/j.1468-2958.2002.tb00813.x

van Haaften T (2019) Argumentative strategies and stylistic devices. Informal Logic 39(4):301–328

Walton D (2010) The structure of argumentation in health product messages. Argument & Computation 1(3):179–198

Wierzba M, Riegel M, Kocoń J, Miłkowski P, Janz A, Klessa K, Juszczyk K, Konat B, Grimling D, Piasecki M, et al. (2021) Emotion norms for 6000 Polish word meanings with a direct mapping to the Polish wordnet. Behavior Research Methods pp 1–16

Chapter 5
Psychological Models of Emotions

Abstract This chapter introduces foundational psychological theories of basic emotions and examines their relevance for the study of emotional appeals in argumentation. It begins by outlining two dominant approaches: the dimensional model, which conceptualizes emotions along continuous scales of valence and arousal, and the categorical model, which identifies distinct basic emotions such as anger, fear, sadness, joy, and disgust. Special attention is given to the theories of Ekman and Plutchik, whose models emphasize stimulus-response structures, automatic appraisal, and adaptive behavioral functions. These models are presented not as comprehensive accounts of emotional life but as simplified, intersubjective frameworks useful for operationalizing emotional appeals in discourse. The chapter distinguishes between emotions expressed by speakers and emotions elicited in audiences, arguing that the latter are especially relevant for understanding pathos in natural language argumentation. By viewing emotion-eliciting language and argument schemes as stimuli that activate fast, affective responses, the chapter offers a psychological foundation for the interactional model of pathos developed throughout the book. This framework prepares the ground for the computational tools and empirical studies that follow.

Keywords Basic emotions · Categorical models · Dimensional models · Ekman · Plutchik · Emotional appeals · Emotional stimuli · Automatic appraisal · Pathos · Audience emotion · Emotion and Persuasion

5.1 What Is an Emotion?

This chapter introduces foundational psychological models that explain what emotions are, how they emerge, and how they influence human behavior and cognition. It focuses on basic emotions, those thought to be evolutionarily ingrained, universally recognizable, and associated with characteristic physiological and behavioral responses. By presenting these models, particularly the dimensional and categorical frameworks, I aim to offer a psychologically grounded vocabulary for analyzing emotional appeals. The chapter lays conceptual groundwork for later sections of

B. Konat, *Emotional Appeals in Argumentation*, Argumentation Library 47,
https://doi.org/10.1007/978-3-032-06332-8_5

the book, which examine how emotional language elicits affective responses in audiences and contributes to persuasive effects in natural language discourse.

The two primary frameworks for categorizing emotions are dimensional and categorical. The dimensional approach represents emotional states in a multidimensional space, where valence (ranging from negative to positive) and arousal (ranging from low to high) represent a continuous flow of emotional states (Wundt 1896; Mehrabian and Russell 1974). Models such as the Circumplex Model of Emotions (Russell 1980) and the Network Model (Lange and Zickfeld 2021) are the newer developments of this approach.

By comparison, the categorical approach identifies several distinct basic emotions, such as anger, disgust, fear, sadness, anticipation, happiness, surprise, and trust, that form the foundation for more complex emotional states (Ekman 1999; Plutchik 2001, 2003). Each theory suggests different numbers and types of these basic emotions, with the models of Paul Ekman and Robert Plutchik among the most widely recognized. Emotions are also an important component in memory (Doerksen and Shimamura 2001), cognition (Joormann et al. 2010), and decision-making (Damasio 1994). This approach is particularly useful in argumentation, where understanding the role of these fundamental emotions can help clarify how emotional appeals influence reasoning and decision-making processes.

One of the central theories on which I will focus is Plutchik's chain-of-events model (Plutchik 2001). This model traces how discrete emotions are triggered by specific events and lead to the corresponding behaviors. For example, joy is often the result of gaining something valuable, and this emotional state leads to behaviors aimed at retaining the object or situation that caused it. Fear is elicited by perceived threats, prompting avoidance or escape behaviors. Anger arises when an obstacle is encountered, leading to aggressive behaviors aimed at overcoming the challenge. Sadness follows the loss of a valued object, leading to behaviors that seek recovery or adaptation. Disgust is triggered by exposure to harmful or repulsive stimuli, prompting rejection or distancing behaviors. I will present how this behavioral understanding of emotions provides a valuable tool for analyzing emotional appeals such as fear appeals and arguments invoking positive consequences.

Wharton and de Saussure (2023) begin their chapter by citing Fehr and Russell's observation that "Everyone knows what an emotion is until asked to give a definition". This remark captures the inherently subjective nature of emotions, which aligns with the concept of qualia (Jackson 1982). By "qualia", I refer to the subjective first-person experiences accompanying perception. For instance, consider the thought experiment of the color specialist who has lived her entire life in a black-and-white world. Although she has never seen the color red, she has comprehensive knowledge about its physical properties and the neural processes involved in perceiving red. Despite this, she lacks direct access to the qualia of red, the subjective experience of seeing the color. This thought experiment shows the difference between knowledge about an experience and the experience itself.

In referring to emotions as involving a qualia-like aspect, I focus on the subjective, first-person dimension of emotional experience, the "what it feels like" aspect. Although emotions also encompass cognitive, physiological, and behavioral components, my use of qualia here is limited to highlighting the distinction between subjective experience and intersubjective or observable models. In this chapter, I adopt an approach analogous to Mary's perspective: I set aside the subjective, qualia-like perspective and instead focus on aspects of emotion that are intersubjective and observable. This mirrors Fehr and Russell's point about the intuitive nature of emotions: While we all have an experiential understanding of them, articulating a precise definition is challenging. Most of us experience emotions, but we may not fully comprehend their underlying mechanisms.

Here, my aim is to present a knowledge-based perspective on emotions that I find particularly useful for argumentation studies, especially in analyzing pathos, the emotional appeal. Consequently, I will omit extensive discussions on subjective, cross-cultural, or therapeutic perspectives on emotions. Instead, I will focus on simplified models that conceptualize emotions in terms of stimuli and reactions, offering tools relevant for practical applications in argumentation.

A full historical review of emotions lies beyond the scope of this chapter, as Wharton and de Saussure have already addressed it thoroughly in their work. Although my approach may not capture the depth of other perspectives, it provides a framework that I believe is instrumental for introducing emotional lexicons and their applications, which will be discussed in subsequent chapters. Starting with Ekman's (1999) perspective emotions evolved with adaptive value in dealing with fundamental life tasks, they share common characteristics: they are all rapid onset, short duration, unbidden occurrence, and automatic appraisal. Emotions often begin so rapidly that individuals may not immediately recognize their onset, especially in situations requiring quick reactions—Ekman brings example of fear appearing when we see an object in periphery of our eyesight that seems to rapidly move towards us. Ekman also argues that emotions are brief and cannot last long—in his interpretation if research subjects were reporting longer lasting emotions, these were in fact just series of emotions. This treatment of emotions as short lasting allows him to distinguish them from moods and other affective states—which last longer even if they are saturated with emotions (like dysphoria with sadness).

Ekman uses the term antecedent events to describe situations or contexts that elicit emotions. These events can be understood as the broader conditions that trigger an emotional response, which includes what we might also call a stimulus. Although a stimulus often implies a specific immediate trigger, such as a loud noise or a fast-moving object, antecedent events may involve more complex or indirect cues that trigger emotions. For instance, both sudden noise and the realization of a betrayal are examples of antecedent events, but they differ in immediacy and complexity. This distinction shows that emotions can arise from a range of triggers, from direct stimuli to broader situational contexts.

Ekman (1999) identifies the automatic appraisal mechanism as a characteristic of emotion closely related to the persuasive process. It is a process by which emotions evaluate stimuli as significant before conscious thought. This mechanism enables rapid, often unconscious, reactions to stimuli that elicit specific emotions. Generally, appraisal occurs quickly, but Ekman noted that it can sometimes be slower, involving more deliberate and conscious thought. The speed and nature of appraisal can be shaped by both evolutionary (phylogenetic) and individual (ontogenetic) factors. For example, certain stimuli, such as the sight of a snake, universally invoke fear, while others vary according to cultural context or personal life history, particularly the emotions associated with the first encounter of an individual with a given stimulus. In terms of emotional appeals in argumentation, this automatic and rapid assessment plays a role in shaping how people evaluate arguments, categorizing their elements as desirable or undesirable, as something to approach or avoid. This concept is central to Walton's (1992) position (explored in Part I) that emotional appeals can lead to fallacies if a weak argument is perceived as stronger than it truly is. Since appraisal interacts with cognitive processes, it influences persuasiveness; however, because this assessment happens quickly and often unconsciously, it can facilitate manipulation or the strategic use of emotional appeal. This effect aligns with the Elaboration Likelihood Model (Petty et al. 1981), which proposes two distinct paths to process persuasive messages, a topic that we will examine in more detail in the following chapters.

Another characteristic of emotion that distinguishes it from other affective and cognitive processes is what Ekman calls an unbidden occurrence. This concept builds on the previously mentioned ideas of rapid onset and automatic appraisal, as emotions generally arise spontaneously in response to stimuli, without our control. Often emotions appear even when we might prefer to avoid them. There are a few exceptions where we can attempt to induce emotions in ourselves, such as choosing to attend a stand-up comedy show or a melodramatic movie. However, in most cases, we have limited control over when and how emotions will emerge. This lack of control can also be exploited in manipulative content. A skilled rhetorician, for example, can use emotional appeals to evoke and modulate the emotions of the audience even within a single speech, guiding responses through strategically timed emotional triggers. This again, aligns with classic understanding of pathos, as elaborated in the first chapter of this book. For Aristotle, pathos was something by necessity caused by external force, something that is experienced rather than actively performed.

What then *is* an emotion? The defining characteristics of emotions: rapid onset, short duration, unbidden occurrence, automatic appraisal, distinguish them from other affective states. These features highlight the adaptive role emotions play in helping people respond quickly to significant stimuli, often before conscious awareness is fully registered. Emotions act as immediate, involuntary responses to external or internal events, preparing us to react in ways that may be beneficial for survival or social functioning. In the next section, we will explore how these shared features contrast with those of specific categorical emotions, following Ekman's argument for the existence of universal basic emotions.

5.2 Categorical Emotions

The defining characteristics of emotions, such as rapid onset, short duration, and automatic appraisal, distinguish them from other affective states, as discussed earlier in this chapter. However, additional features such as signaling functions, physiological responses, and specific antecedent events are unique to distinct categorical emotions. Ekman (1999) considers these distinctions to be central to his argument for the existence of basic emotions. Keeping within the framework of emotions as reactions to specific stimuli (or antecedent events), we will now explore the reasoning of Ekman to identify basic emotions as distinct categories. I share Ekman's view that the positive-negative scale (or pleasant-unpleasant) is often inadequate, particularly when it comes to emotional appeals in argumentation. We frequently refer to specific emotions, such as "appeal to pity" or "fear appeal", but what exactly do we mean by these terms? In the argumentation literature, these labels are often used in a general sense, reflecting common language rather than technical precision. Walton (1992) himself points out that logical and prescriptive discussions do not necessarily require adopting the full theoretical and empirical framework of social science. Although a comprehensive overview of the psychology of emotions would be impractical, given the thousands of volumes dedicated to the topic, some theoretical foundation is valuable.

Ekman identifies six basic emotions: anger, fear, sadness, enjoyment (also referred to as joy or happiness), disgust, and surprise. By "basic", Ekman means emotions that have evolved due to their adaptive value in the face of fundamental life challenges and that share universal signals, physiological responses, and typical antecedent events across cultures. However, these basic emotions are not rigid, isolated categories. Instead, Ekman describes them as "emotion families", which means that each basic emotion includes a range of related emotional states that share a common theme but vary in intensity and expression depending on individual and contextual factors. For example, the family of anger can include related states like irritation or fury, each with distinct signals but sharing the same core characteristics of anger. Ekman's concept of emotion families allows flexibility within basic emotions, reflecting both universal characteristics and individual differences.

Table 5.1 presents the summary of "feeling states" (that is, analogous to human emotional states observed in animals other than humans) as listed in Plutchik's work (Plutchik 1980, 2001). Let this be the start of our overview of categorical emotions understood in terms of stimulus and reactions. The table outlines a series of typical stimulus events and their corresponding emotional outcomes. Each stimulus event triggers a cognitive assessment, essentially a mental interpretation that characterizes the nature of the event. I believe that Plutchik's work complements Ekman's theory by offering a more detailed account of how emotions serve as reactions to specific stimuli, not only shaping behavior but also achieving functional outcomes. This view supports the concept of basic emotions as distinct categories that arise from fundamental, evolutionarily developed processes. Let us now analyze each emotion separately.

Anger arises when an individual perceives an obstacle or injustice that blocks a goal, a need, or an expectation. This stimulus triggers an appraisal that categorizes the obstacle as a threat or offense, activating physiological arousal such as increased heart rate, muscle tension, and heightened alertness. Anger's primary behavioral response is confrontation, often manifesting in assertive speech or aggressive

Table 5.1 The table shows the relationship between different stimulus events, cognitive appraisals, feeling states, overt behaviors, and their effects

Stimulus event	Cognition	Feeling state	Overt behavior	Effect
Threat	"danger"	Fear	Escape	Safety
Obstacle	"enemy"	Anger	Attack	Destroy obstacle
Gain of valued object	"possess"	Joy	Retain or repeat	Gain resources
Loss of valued object	"abandonment"	Sadness	Cry	Reattach to lost object
Member of one's group	"friend"	Acceptance	Groom	Mutual support
Unpalatable object	"poison"	Disgust	Vomit	Eject poison
New territory	"examine"	Expectation	Map	Knowledge of territory
Unexpected event	"what is it?"	Surprise	Stop	Gain time to orient

actions, depending on intensity. These responses help mobilize energy toward overcoming or eliminating the perceived threat, with the ultimate goal of restoring control or balance. Ekman's "anger family" recognizes that this emotion varies in intensity, from mild irritation to intense rage, with a distinct set of facial expressions, such as furrowed eyebrows and tight lips, that universally communicate the presence of anger to others (Ekman 1999). This is also reflected in Plutchik's concept of a wheel of emotion, where instead of facial expression more attention is towards the subjective feeling.

Fear emerges in response to a perceived threat, typically one that suggests immediate physical or psychological harm. Such stimuli are appraised as dangerous, initiating the "fight or flight" response, which includes heightened heart rate, sweating, and, in some cases, freezing. These physiological changes prepare the individual for rapid escape or defensive actions, oriented toward self-preservation. Fear facial expressions, such as wide eyes, raised eyebrows, and a tense mouth, serve to increase sensory awareness, allowing individuals to detect and evaluate potential threats more effectively. Ekman's work (1999) also notes that fear can be anticipatory, such as fear preceding an uncertain event, but in either case, fear functions to prepare people for danger, increasing survival chances. Again in Plutchik's model this category covers emotions from apprehension, trough fear, up to terror—depending on intensity.

Sadness is typically triggered by loss, whether of a valued object, a personal relationship, or a cherished goal. The appraisal process categorizes the loss as abandonment or failure, leading to emotional and physiological responses that slow the body's energy and reduce motivation. The typical behavioral response to sadness involves withdrawal and reflection, although it can also prompt a person to seek comfort or support from others. Ekman describes sadness as part of an "emotion family" that includes various intensities, from disappointment to profound grief. Expressions associated with sadness, such as drooping eyelids, frowning, and tears, are universally recognized and often function as social signals, indicating a need for empathy and support within communities. In Plutchik's model this category starts with pensiveness through sadness up to grief.

Enjoyment or joy is an emotion that arises from positive events, such as achieving a goal or experiencing something pleasurable. The stimulus is appraised as beneficial, reinforcing the desirability of the event. Joy produces physiological responses associated with elevated energy levels and the release of neurotransmitters like dopamine. Behaviorally, joy encourages individuals to sustain or repeat the rewarding experience, expressed through smiles, laughter, and increased social engagement. Ekman notes that enjoyment includes a family of related states, ranging from mild contentment to intense euphoria. Among the facial expressions of joy, the "Duchenne smile" is characterized by activation of the eye muscles along with the mouth, distinguishing genuine joy from polite or social smiles. This emotion strengthens social bonds and motivates continued pursuit of rewarding activities. In Plutchik's model, this category encompasses serenity, joy, up to ecstasy.

Disgust is typically triggered by stimuli that are perceived as offensive or potentially contaminating, often relating to taste, smell, or moral sensibilities. This appraisal identifies the stimulus as harmful, activating an aversive response aimed at expelling or distancing itself from the offending object. Physiologically, disgust often includes sensations of nausea and muscle contractions associated with rejection. Behaviorally, disgust results in actions such as turning away, covering the nose, or making sounds of distaste. Ekman notes that disgust serves an adaptive function by protecting individuals from potential toxins or moral violations. The distinct facial expression associated with disgust, a wrinkled nose and a raised upper lip, is universally understood and functions to communicate aversion, aiding in social cooperation by signaling to others the presence of something offensive or harmful. For Plutchik, interestingly, the milder version of disgust is boredom, whereas the strongest is loathing.

Surprise arises in response to an unexpected or novel event that disrupts expectations. The stimulus is appraised based on its novelty, which triggers a rapid orientation response aimed at gathering additional information. Physiologically, surprise involves a brief increase in alertness, with facial expressions such as raised eyebrows, widened eyes, and an open mouth. These expressions serve to enhance sensory intake and prepare the individual for rapid evaluation of the unexpected event. Surprise is generally the shortest lasting of Ekman's basic emotions, as it quickly transitions to another emotion once the appraisal process categorizes the nature of the stimulus. The universal expression of surprise signals to others that attention is required, helping groups respond more quickly to sudden changes in the environment. Surprise is a difficult emotion, with even less clear boundaries than the remaining five, and hence it is often excluded in research. This has been acknowledged by Ekman himself, and we will go back to this topic when looking at empirical results in later chapters of this book, which confirm the different nature of surprise. For Plutchik this emotion in its lightest intensity is amazement and in stronger form—a distraction. As we shall later see, the quick fading of this emotion and the fact that it originally is neither positive nor negative strongly influences how it is perceived by both human subjects in experimental setting but also by automated models in computational linguistics and AI.

5.3 Psychological Models of Emotion and Argumentation Theory

The key takeaway from these psychological approaches is that emotion is a reaction, a useful, evolutionarily grounded response to stimuli. This stimulus can be real, like a dangerous object approaching or the face of a loved one, but it can also be, in a cultural setting, a representation of such an object, either iconic (as in a movie or photograph) or symbolic, as in language. Thus, certain words or argument schemes can act as stimuli that elicit emotions. This psychological view of words as stimuli is closely aligned with Aristotle's concept of pathos (as discussed in Part I) as an attempt to evoke emotions in the audience through speech. This is what I mean by emotional appeals "in" argumentation—I do not claim that pathos is part of logos in the conceptual sense, just that emotional appeals are often interwined in natural language argumentation in the same linguistic elements, providing, however, a different level of persuasive strategies.

This alignment is why I have chosen to apply these psychological tools to an interactional model of pathos in which the speaker, aiming to increase persuasiveness, deliberately selects words or constructions to elicit emotions in the audience. This approach not only aligns with the Aristotelian view of pathos but also extends it substantially. It parallels Macagno and Walton's concept of 'emotive words' (also discussed in Part I), with its focus on words that evoke emotions. It also encompasses argument schemes such as *ad baculum* or *ad misericordiam*, where the focus is not on expressing emotions but on appealing to them. In this way, I believe these psychological concepts provide a framework for operationalizing the concept of emotional appeal, which is central to this book. In the next chapter, I will introduce the tool of affective lexicons, where I refer to these as 'emotion-eliciting words.' However, it is important to note that this approach contrasts with other perspectives that focus on expressed emotions, such as those of Plantin, Cigada, and others. Although I agree that the speaker's own expressed emotions play a significant role in persuasion, they require a separate approach and fall outside the scope of this book.

In other words, the key distinction I propose here is between eliciting emotions in the audience and expressing emotions by the speaker. Traditional approaches in emotional language analysis often emphasize the speaker's displayed emotions. In contrast, my focus is on emotion-eliciting language and argument schemes that influence audience emotions without necessarily revealing the speaker's own feelings. For example, in an appeal to pity (ad misericordiam), the speaker does not need to appear personally affected but instead selects words that elicit sympathy directly in the audience. Similarly, purposefully using words such as "catastrophy" or "danger" in one's argument (where a word such as "problem" would be sufficient), the speaker is not necessarily expressing their own emotion, but instead is attempting at eliciting fear in the audience. In doing so, this book aims to offer a new perspective and practical methodology for analysing real world argumentation.

Given the quick onset and short duration of emotional responses (despite some debate on this point), we can assume a fast-acting role of pathos in persuasion. When pathos is combined with logos, it not only interacts with the audience's existing mood but also rapidly creates the emotional setting in which arguments are processed. Although the immediate effect of pathos may be brief, it serves as a backdrop that influences how arguments are appraised, remembered, and judged. In this way, pathos can have a lasting, indirect impact on the persuasiveness of an argument.

In this chapter, I have outlined psychological models of basic emotions with the aim of providing a conceptual foundation for analyzing emotional appeals in argumentation. By focusing on the adaptive, stimulus-response nature of emotions, particularly as conceptualized by Ekman and Plutchik, I have emphasized that emotions can be understood as rapid, automatic reactions to salient stimuli, whether directly perceived or symbolically represented. This view enables a practical distinction between emotions expressed by speakers and those elicited in audiences. It is the latter (emotions experienced by the audience in response to particular linguistic cues) that is central to this book's account of pathos.

Understanding emotions as reactions to stimuli allows us to treat words and argumentative structures as potentially eliciting emotions. This is a key for the development of an interactional model of pathos, MIPA, introduced earlier. Rather than focusing on introspective or culturally variable definitions of emotion, I have adopted simplified, intersubjective models that lend themselves to empirical study and computational implementation.

The next step in this framework is to explore how emotional stimuli can be systematically identified in discourse. In the following chapter, I introduce the selection of affective lexicons, curated sets of words annotated for emotional properties such as valence, arousal, and discrete emotion categories. These lexicons offer an operational tool for identifying and analyzing emotion-eliciting language in texts. In this sense, the next chapter bridges conceptual understanding with empirical application, moving from theoretical distinctions to concrete tools for analyzing pathos in natural language argumentation.

References

Damasio AR (1994) Descartes' Error: Emotion, Reason, and the Human Brain. G.P. Putnam, New York

Doerksen S, Shimamura AP (2001) Source memory enhancement for emotional words. Emotion 1(1):5–11, https://doi.org/10.1037/1528-3542.1.1.5

Ekman P (1999) Basic emotions. In: Dalgleish T, Power M (eds) Handbook of Cognition and Emotion, Wiley Online Library, pp 45–60

Jackson F (1982) Epiphenomenal qualia. Philosophical Quarterly 32(127):127–136, https://doi.org/10.2307/2960077

Joormann J, Yoon KL, Siemer M (2010) Cognition and emotion regulation. In: Kring AM, Sloan DM (eds) Emotion regulation and psychopathology: A transdiagnostic approach to etiology and treatment, The Guilford Press, pp 174–203

Lange J, Zickfeld JH (2021) Emotions as overlapping causal networks of emotion components: Implications and methodological approaches. Emotion Review 13(2):157–167, https://doi.org/10.1177/1754073920986182

Mehrabian A, Russell JA (1974) An Approach to Environmental Psychology. MIT Press, Cambridge, MA

Petty RE, Cacioppo JT, Goldman R (1981) Personal involvement as a determinant of argument-based persuasion. Journal of Personality and Social Psychology 41(5):847–855, https://doi.org/10.1037/0022-3514.41.5.847

Plutchik R (1980) Chapter 1 - a general psychoevolutionary theory of emotion. In: Plutchik R, Kellerman H (eds) Theories of Emotion, Academic Press, pp 3–33, https://doi.org/10.1016/B978-0-12-558701-3.50007-7, https://www.sciencedirect.com/science/article/pii/B9780125587013500077

Plutchik R (2001) The nature of emotions: Human emotions have deep evolutionary roots, a fact that may explain their complexity and provide tools for clinical practice. American scientist 89(4):344–350

Plutchik R (2003) Emotions and life: Perspectives from psychology, biology, and evolution. American Psychological Association

Russell JA (1980) A circumplex model of affect. Journal of personality and social psychology 39(6):1161

Walton D (1992) The Place of Emotion in Argument. The Pennsylvania State University Press, University Park, PA

Wharton T, de Saussure L (2023) Pragmatics and Emotion. Cambridge University Press

Wundt W (1896) Gundriss der Psychologie [Outlines of Psychology]. Entgelmann, Leipzig, Germany

Chapter 6
Empirical Applications

Abstract This chapter presents empirical methodologies for studying emotional appeals in argumentation, integrating psychological models with experimental approaches. It begins by introducing affective lexicons based on dimensional and categorical models of emotion, including large-scale datasets across multiple languages. These lexicons are shown to reflect cross-linguistic patterns in emotional word perception and to serve as tools for identifying potential emotional appeals in discourse. Building on the interactional model of pathos, the chapter then explores how such emotion-eliciting words function within arguments. The second part of the chapter focuses on experimental research. It reviews recent psycholinguistic and cognitive studies investigating emotional arguments, including the use of physiological and behavioral measures. Finally, the chapter presents two experimental studies conducted by the author's team: one measuring the effects of personal involvement, mood, and emotional appeals on Perceived Argument Strength (PAS), and another analyzing emotional appeals in student debates. Findings confirm that while emotional appeals, especially positive ones, can enhance perceived strength, audience-related factors such as involvement and mood are stronger predictors.

Keywords Emotional appeals · Affective lexicons · Basic emotions · Argument strength · Audience effects · Elaboration Likelihood Model · Perceived Argument Strength · Experimental pragmatics

6.1 Lexicons of Emotion-Eliciting Words

6.1.1 Dimensional Models of Emotion in Lexicons

Building on the psychological models of emotion introduced in the previous chapter, this chapter turns to empirical applications, demonstrating how emotional appeals in argumentation can be operationalized, measured, and analyzed using affective lexicons and experimental methods.

Starting with Osgood et al.'s (1957) dimensional model, the chapter demonstrates how emotional models can be empirically linked to specific words, allowing the

B. Konat, *Emotional Appeals in Argumentation*, Argumentation Library 47,
https://doi.org/10.1007/978-3-032-06332-8_6

creation of systematic emotion lexicons. One well-known example is the Affective Norms for English Words (ANEW; Bradley and Lang 1999), developed to help psychologists select words for experiments to control for unintended emotional responses. This lexicon was created using a standardized methodology where participants rated valence and arousal on fixed scales. Later, other significant data sets for English were introduced (Stevenson et al. 2007; Warriner et al. 2013). For nonverbal stimuli, such as images and videos, emotion datasets tend to be more universally applicable across cultures, such as the International Affective Picture System (IAPS) (Lang et al. 2005). The lexicons of verbal stimuli (words, sentences, etc.) are often language- and culture-specific, creating the need for cultural adaptation. Several emotion lexicons have been developed for languages such as French (Monnier and Syssau 2014), European Spanish (Guasch et al. 2016), Argentinian Spanish (Sarli and Justel 2022), Italian (Montefinese et al. 2014), European Portugese (Soares et al. 2012), Polish (Imbir 2016; Wierzba et al. 2021) and multiple other languages. Alternative methodologies, such as crowd-sourcing (Mohammad and Turney 2010, 2013), have also been employed, as in a recent study that I contributed to, which combined in-lab and crowd-sourced methods (Wierzba et al. 2021).

Going back to the origins of this methodology, the Affective Norms for English Words (ANEW) is one of the earliest and most widely used affective lexicons, developed to provide standardized emotional ratings for English words. Introduced by Bradley and Lang (1999), ANEW was developed with the intention of providing a normative reference point that would help researchers compare the results in the studies, aiming to ensure greater consistency and support replication in the area of studies on emotion and attention. ANEW is based on a dimensional model based on Osgood et al. (1957), in which three primary dimensions were identified: affective valence (pleasant to unpleasant), arousal (calm to excited), and dominance (control).

In terms of affective lexicons, let us consider selected examples from the Spanish lexicon: "emoción" (emotion), "relajación" (relaxation), "enfermedad" (disease), "desánimo" (despondency), "riesgo" (risk), "perezoso" (lazy) (Guasch et al. 2016). If we were to plot these words on a chart with arousal on the vertical axis (ranging from low to high) and valence on the horizontal axis (from negative to positive), we would observe a characteristic U-shaped distribution. Words with very positive valence, such as "joy," or very negative valence, such as "disease," are positioned at the outer ends of this U-shape, reflecting their strong emotional impact. In contrast, words with neutral valence, such as "lazy," typically exhibit lower arousal, placing them within the central dip of the U. This distribution, observed consistently across multiple languages, suggests a simple regularity: words perceived as evoking positive or negative emotions are often seen as more emotionally moving and correlate with higher arousal, while neutral words are generally perceived as less arousing, creating the dip in the U-shape.

However, some words are outliers to this pattern. Take, for example, the word "risk." In languages such as Polish and Spanish, "risk" defies the usual trend, showing a combination of near-neutral valence and high arousal. This ambivalence makes "risk" appear arousing despite its neutral valence. This paradox can be explained by a high standard deviation in responses to "risk", indicating that some respondents experience strong positive emotions and excitement, while others feel strong negative emotions accompanied by arousal. This results in an artificial average of neutral valence, which is actually a blend of opposing reactions. A similar situation can be observed in ANEW, where the word "champion" received a high valence rating, indicating a consistent positive response, while the word "rifle" seems to have a neutral valence; however, the presence of a higher standard deviation shows that the participants experienced more varied emotions, some positive and some negative. This shows the limitations of affective lexicons, which sometimes contain such artifacts.

Remarkably, the U-shaped distribution had been found across multiple languages and language families, even when analyzing distinct parts of the lexicon. This consistency illustrates the strength of this method. To explore cultural differences and similarities, we can look at the words that occupy the upper right extremity of the U-shape: often, it is simply the word "joy" or a similar term. In the affective lexicons of English, Spanish and Polish, we find that "joy" (English), "emoción" (Spanish), and "radość" (Polish)—respectively—occupy areas of high valence (positivity) and high arousal, showing a cross-linguistic association with intense positive emotions. However, there are also differences: In Spanish, it is "emoción" rather than the more literal name for "joy". In Polish, this position is occupied by "szczęśliwy" (translated as "a happy man") rather than the straightforward translation of "joy." These examples show both how certain elements are universal across cultures and how others differ.

Although cultural adaptations of affective lexicons are needed, and we should be cautious of Western-centric biases, there is considerable empirical evidence that valence and arousal are useful universal measures for understanding emotional responses across languages and cultures. What is also interesting is another universality—remember that words are presented to participants without any linguistic context, therefore the consistent affective ratings suggest that certain emotional associations are deeply rooted in shared human cognition.

Still, one challenge with emotion lexicons is the subjective and context-dependent nature of emotions. Reactions to specific words can vary greatly depending on context and individual differences. However, lexicon developers use controls and statistical techniques to ensure the reliability of the data. Usually, each word is rated by about twenty-five different participants, ensuring that average values and standard deviation can represent a reliable sample. Although emotional reactions are subjective, lexicons provide useful generalizations about how populations respond to words in the absence of context. However, these generalizations should be interpreted with caution. Recent advances in computational linguistics, including the work of my team (Gajewska et al. 2024), have addressed these challenges by modeling individual emotional responses in more detail, which I will discuss in further detail in Part III.

6.1.2 Models of Categorical Emotions in Lexicons

Affective lexicons are not only valuable for identifying the valence and arousal levels associated with words, but they also allow for a deeper analysis by incorporating models based on basic categorical emotions. In the previous sections, I introduced Robert Plutchik's concept of eight fundamental emotions (anger, disgust, fear, sadness, anticipation, happiness, surprise, and trust) as described in his work. Now, I would like to briefly discuss an emotive lexicon based on this model, in which development I took part (Wierzba et al. 2021).

We created emotion norms for 6000 Polish words (specifically for meanings mapped to the Polish WordNet, although we will set that aspect aside for now). A key strength of our study lies in its scale: we had more than 500 participants in the lab portion and more than 20,000 online. This large sample allowed us to confirm that annotations were generally consistent across demographic groups, including both female and male participants as well as younger and older groups. While demographic factors influenced the responses to some extent, the overall patterns remained stable.

Our methodology was consistent with standard practices for similar lexicons. Participants were encouraged to provide their immediate reactions to words, although no time limit was imposed. They viewed each word and a related phrase at the top of the screen and then rated each word on valence, arousal, and basic emotions using sliders. For each emotion, they could move the slider between 0 and 4, which allowed them to indicate multiple emotions for each word.

To ensure that all basic emotions were well represented in our dataset, we applied specific selection criteria. Each word was rated for the intensity of each of the basic emotions, meaning a word could be linked with a single dominant emotion (such as happiness) or associated with several emotions (like anger and disgust), or none at all. Out of an initial pool of 30,080 word meanings, we selected 6000 in total: 5000 with a single dominant emotion (divided evenly across the eight emotion categories) and 1000 neutral words. The threshold values for inclusion were established using a basic genetic algorithm.

The resulting lexicon,[1] shows how words can evoke various emotions, building on previous lexicons that primarily focused on valence and arousal. Words with a higher positive valence, such as happy ("zadowolony"), smiling ("uśmiechnięty"), face ("buzia"), very nice ("przesympatyczny"), delightfully ("rozkosznie"), picnic ("piknik"), adorable ("przesłodki"), sun ("słońce"), to joke ("żartować"), honored ("wyróżniony"), and wonderfully ("cudownie"), are rated highest for happiness, often in combination with strong anticipation scores. Unsurprisingly, they all belong to the class of happiness. However, these words also receive slight but non-zero ratings for emotions such as anger or disgust, reflecting individual differences among participants who perceived subtle traces of these emotions.

[1] Lexicon publicly available at https://osf.io/f79bj/.

This again demonstrates how lexicons reveal general trends while also capturing individual variability.

On the opposite end of the valence scale, we find words with the lowest ratings, such as slimy ("oślizgły"), filthy ("syfiasto"), fecal ("fekalny"), deceased ("zmarła"), unhygienic ("niehigieniczny"), ulcerated ("owrzodziały"), stinker ("śmierdziel"), funeral ("pogrzeb"), extremely unpleasant ("arcyniemiły"), and smelly ("śmierdzęcy"). In terms of emotional classification, these words evoke a wider range of negative emotions: sadness for deceased ("zmarła") and funeral ("pogrzeb"), anger for extremely unpleasant ("arcyniemiły"), and disgust for the remaining words.

These words are relatively clear-cut and likely to elicit negative emotions across different cultural backgrounds. But what does it really mean for a word to be assigned to a specific emotional category in a categorical lexicon? This question becomes especially relevant when we consider less straightforward cases.

If we examine standard deviation, such as with the word risk ("ryzyko") in previous examples, to identify words with the least consensus among respondents, the first ten are as follows: pornographic ("pornograficzny"), fecal ("fekalny"), scary ("straszny"), liver ("wątróbka", in the context of eating it), nasty ("paskudny"), ceremony ("ceremonia"), noisy ("hałaśliwy"), rotten ("zgniły"), insurgent ("powstaniec"), and depressing ("dołowy"). These words span different emotional categories, mostly negative: disgust for pornographic ("pornograficzny"), fecal ("fekalny"), nasty ("paskudny"), noisy ("hałaśliwy"), and rotten ("zgniły"), fear for scary ("straszny"), sadness for ceremony ("ceremonia"), insurgent ("powstaniec"), and depressing ("dołowy"), and anger for noisy ("hałaśliwy"). This illustrates how reactions can vary significantly depending on individual and cultural factors. For example, some readers may respond to the idea of eating liver with anticipation and happiness, while others may find it utterly repulsive. In some readers, the concept of "ceremony" brings joyous memories of weddings and parties, whereas for others the first association evokes sadness, probably because this word is associated with funerals. Our respondents were similarly divided on certain words. Therefore, when using lexicons, it is essential to remember that the association of a word with a particular emotion category indicates that, on average, it tends to evoke this emotion more often. However, other emotions may also be elicited and individual reactions can vary widely.

In summary, the model of Robert Plutchik that describes eight basic emotions, anger, disgust, fear, sadness, anticipation, happiness, surprise, and trust, can be applied to lexicons. Emotional lexicons use this and similar models to categorize emotions associated with specific words. They are a useful tool and source of information about the emotion that a particular word is likely to evoke and are based on extensive large-scale psychological studies. They show that, contrary to the idea that everyone's emotional response is entirely unique and unpredictable, there are general trends in how people react to certain words, even though there are still individual variations.

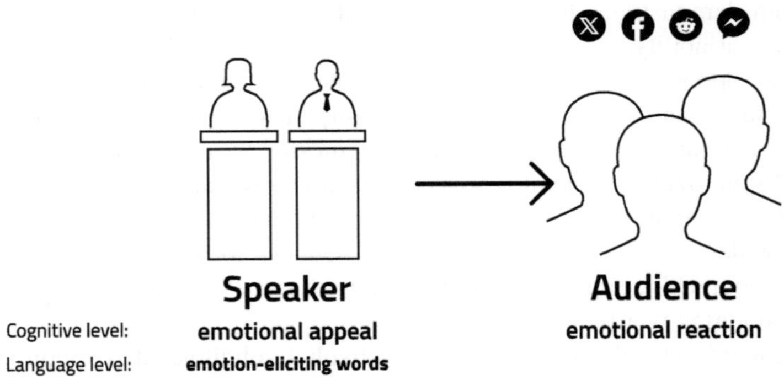

Fig. 6.1 Model of Interactional Pathos in Argumentation (MIPA): emotion-eliciting words in speaker's arguments

Going back to the description of the Model of Interactional Pathos in Argumentation (MIPA), we said that affective lexicons are build by collecting responses to words—participants are instructed to mark, without spending too much time on each word, their first reaction by pointing on a visual scale in each of primary dimensions described earlier. This is underlaid by the psychological concept of treating each word as a stimuli, something that acts on a human and causes reaction. We can see how this psychological approach to word as a stimuli fits into the concept of pathos as an appeal to emotion of the audience, as presented in MIPA. Words such as "disease" or "risk" do not express any emotions; however, they can elicit one when heard. Thus, emotion-eliciting words listed in affective lexicons can be used by speakers who can include them inside the language of their arguments as a linguistic means to perform emotional appeal.

Figure 6.1 presents the Model of Interactional Pathos in Argumentation (MIPA) introduced in Part I, as applied to the use of affective lexicons. On the speaker's side, we see that at the cognitive level, they aim to make an emotional appeal. Previously, we analyzed how this can be achieved using specific argument schemes. Now, considering the structure and methodology of emotional lexicons, it is clear that the words within these lexicons can be used, and often are, by speakers to elicit emotions in their audience. In Part III of this book, I will demonstrate how, by using affective lexicons and simple computational methods, researchers in argumentation and persuasion can automatically observe patterns in the use of emotion-eliciting words within natural arguments.

This focus on emotion-eliciting words contrasts with some approaches in argumentation theory and discourse analysis. For example, Cigada (2019) emphasizes "emotion terms," explores how speakers express emotions such as appreciation or tension, while Plantin (2019) analyzes emotional expressions such as exclamations from the speaker's perspective. Also, classical sentiment analysis methods, which I will say more about in Chap. 8, focus mainly on emotional expression. On the other hand, psychological lexicons align more closely with Macagno and Walton

(2014) who follow the Aristotelian conception of pathos, focusing on emotions elicited in the audience rather than expressed by the speaker. My work continues this tradition, enhancing it with computational aspect by applying psychological lexicons to identify emotional appeals in natural language argumentation, which I will expand on in Chap. 8.

6.2 Experimental Work in Argumentation Theory

The psychological basis of affective lexicons, treating words as stimuli that elicit emotional reactions, naturally invites empirical investigation. If such words play a role in emotional appeal, as argued above, we can test the persuasive impact of language experimentally. The following section reviews how experimental methods have been used in argumentation theory to study emotional appeals, setting the stage for my own empirical work.

In this section I will be focusing on experiments that aim to capture the persuasive power of emotional appeals in argumentation. The previous chapters introduced the psychological concepts of basic emotions and lexicons of emotive words; here, we focus on another recent adoption from psychology that has gained traction in argumentation theory, experiments with human subjects. Alongside a review of existing experimental studies, I present two of my own experiments, previously published, that investigate the perceived strength of emotional argumentation. Argumentation theory and psychology have long provided each other with support in the form of interesting empirical questions and practical methods for approaching them. Pragma-dialecticians have long studied how the norms of reasonableness can be accepted by laypeople (van Eemeren et al. 2009). Argumentation journals have increasingly opened their scope to experimental work in recent years (Benlamine et al. 2015, 2017; Villata et al. 2017), while psychologists have also begun to explore the empirical study of argumentation effectiveness (Blanchette 2006; Walkowiak and Imbir 2018). Even abstract argumentation (Dung 1995) can be approached experimentally, and interestingly, it has been explored in numerous studies (Rahwan et al. 2010; Cerutti et al. 2014; Polberg and Hunter 2018).

An interesting area resulting in multiple experimental approaches is the intersection between argumentation and reasoning, which naturally brings the cognitive layer. Jos Hornikx (Hornikx and Hahn 2012) calls for an integrated psychology of argumentation, pointing out that argumentation-related psychological studies are still sparse and conducted by teams coming from different fields: either from experimental linguistic pragmatics, developmental psychology, or education and then entering the field of argumentation, or from within the argumentation community itself—yet the picture remains fragmented. The special issue of *Thinking and Reasoning* journal devoted to the topic was one step towards completing this goal. One particularly interesting paper in this collection is Heit and Rotello (2012), which studies how the length of an argument influences persuasiveness. The setup is quite similar to our study presented below: a factor related to the internal structure of the argumentative message (length) is analyzed in the context of two conditions (weak and strong arguments), and what is tested on human subjects is how these influence their perception of how persuasive it is. In the following paragraphs, I will draw on this and other studies to illustrate a methodology that connects theoretical questions rooted in argumentation theory with experimental psychological methods to provide empirical evidence.

One major strand of work comes from the intersection of argumentation theory and experimental pragmatics. Experiments have been commonplace in cognitive pragmatics for years, and I would like to highlight some of the methods from the Swiss team led by Steve Oswald at Fribourg University, which often adopts psychological methods to answer theory-based questions from pragmatics and argumentation. In Schumann et al. (2019), they adopt a method similar to ours by taking a problem rooted in argumentation theory, specifically, the conditions under which a straw man fallacy may seem acceptable, and conducting experiments with human subjects to test hypotheses derived from theoretical perspectives. The concept of a fallacy, and the straw man fallacy in particular, is adopted from argumentation theory, whereas the methods (operationalization, the construction of experimental conditions, and statistical methods of data analysis) are taken from linguistics and psychology. Their focus is to test whether linguistic factors, mainly connectives, influence the acceptability of fallacious straw men as sound arguments. Results suggest that using a form of paraphrasing of an opponent's meaning (echoing) can lead to higher acceptability of fallacies due to the illusion of similarity, while using a connective (i.e., a linguistic explicit marker of pragmatic relations) will increase participants' awareness that something has been changed and increase their vigilance towards the possible fallacy. Another publication from the same team (Younis et al. 2023) analyses the theoretically grounded concept of rephrasing (Konat et al. 2016; Koszowy et al. 2022) to empirically test whether people perceive rephrased statements as being more persuasive than non-rephrased statements. The results suggest that rephrasing the statement in the absence of argumentation is less persuasive than an argument itself, but still brings some persuasiveness. In other words, even in the absence of a strong argument, repeating the same statement in different wording can still have a persuasive effect. Both of these experiments from the same team illustrate how concepts rooted in theoretical approaches, such as fallacy, straw man, rephrase, can be enriched with a cognitive layer and tested with psycholinguistic experiments to support or challenge existing concepts.

Another notable example is the work of Francesca Ervas, previously mentioned in the section devoted to metaphors. Her paper published in *Synthese* (Ervas 2019) provides an in-depth analysis of how ignorance participates in the interpretation of metaphors in argumentation. However, what is crucial is that this theoretical contribution provided grounds for the empirical evaluation presented in Ervas (2021), where the team looks at the combination of two factors in the argumentation: the presence of metaphors and the presence of affective framing. Furthermore, they study it in an insightful manner, providing an operationalization of weak arguments, as arguments with an equivocation fallacy. They adopted a standard syllogistic form for the simplicity and comparability of conditions. On the side of the dependent variable, they created a questionnaire with a Likert scale that tests seven factors: understandability, convincingness, emotional appeal, logical relation, ambiguity, belief in the conclusion, and real-world experience. The results suggest that even though metaphors and emotional appeals do not always automatically decrease the ability of logical evaluation, they do, in the authors' interpretation, activate heuristic, quick thinking, and might lead to a loosening of logical rigor. At the same time, as it is with heuristics, they are useful and used on an everyday basis as quick approximations for the best course of prudential actions, yet they entail the risk of becoming fixed biases. Francesca Ervas' team research provides a useful background for our research in that it has all that we also try to provide from a slightly different perspective: deep theoretical insight and solid empirical experimental grounding.

Particularly relevant to our studies presented here is the interesting and unique work resulting from the cooperation of Université Côte d'Azur and the University of Montreal (Benlamine et al. 2015; Villata et al. 2017). The authors applied the concepts of logos, ethos, and pathos as rhetorical strategies and studied various physiological factors to assess the effectiveness of each strategy. The setting of the first study (Benlamine et al. 2015) was particularly interesting, as the participants participated in a debate with a fictive participant who proposed arguments either for or against those proposed by other participants. All of this occurred while they were connected to tools collecting physiological data to detect their cognitive and emotional reactions: a camera to detect facial expressions, an electroencephalogram (EEG) headset, an electrodermal activity (EDA) bracelet sensor, and an eye tracker. The authors hypothesized that logos-based arguments would be processed differently from those using ethos and pathos as rhetorical strategies: first, that logos arguments would solicit one side of the brain, while ethos and pathos would be more diverse; furthermore, that logos arguments would induce a higher workload; and second, that the most effective strategy would be pathos. The results showed that the left-right brain distinction and the level of cognitive engagement were not directly related to the type of rhetorical strategy, but rather to the stance of the participant: usually, both hemispheres were active and the level of engagement was highest in participants who were still undecided at the end of the debate. This suggests that cognitive engagement is highest when we are looking for reasons to accept or reject an argument; however, once we adopt a stance, these levels tend to decrease. For the pathos results, most relevant for our study here, the findings suggest that no participant remained neutral after this rhetorical strategy, meaning that appealing to emotions is effective, though it may also result in stronger opposition to the argumentation presented. Taking all this together, this study shows the methodology of empirical studies in argumentation and rhetoric—the concepts of logos, ethos, and pathos can be operationalized into psychological and even physiological variables, and so can the concept of effectiveness or persuasiveness.

The second work of this team, published slightly later (Villata et al. 2017), adds an interesting layer of analysis by incorporating personality traits, contributing to our understanding of the effects of audience persuasion. As in the previous study,

EEG signals and facial expressions of emotions were collected. First, the emotional layer provided insight into how anger emerges when arguments are challenged, often evolving into disgust if participants fail to convince their opponents. In terms of cognitive dimensions, engagement and workload, they are strongly influenced by emotional states. Specifically, high engagement correlates with strong emotional responses (both positive and negative), while low engagement is associated with emotions like disgust, which also leads to reduced argumentative activity. The cognitive workload decreases when the participants feel angry, suggesting a reduction in the ability to generate new arguments. In contrast, when disgust emerges, the workload increases, indicating that the participants are reconsidering or reconstructing their arguments. A pioneering look at the role of personality traits taken by the authors suggested that neurotic individuals are more likely to experience anger and disgust, while individuals with higher agreeableness or extroversion might be more open to experiencing emotions such as surprise. This study demonstrates the utility of distinct emotional categories in psychology, such as anger and disgust, for examining reactions in argumentative exchanges. By integrating psychological constructs and methodologies, researchers can gain a deeper understanding of the cognitive and psychological dimensions of argumentation. In the following section, I will describe my own experimental study, which investigates how personal involvement, mood, and emotional appeals influence perceived argument strength. This study builds upon the concepts and methodologies discussed above, further demonstrating the value of empirical approaches in argumentation theory.

In this chapter, I have provided an overview of how psycholinguistic methods, traditionally used in psychology, can contribute to argumentation theory by empirically testing claims found in handbooks, such as "emotional arguments can appear stronger than they are" and "emotional appeals are used to win debates." By applying experimental approaches, I demonstrate that factors influencing argument strength, many of which are audience-related and well-studied in psychology, can be more precisely articulated and validated within argumentation theory. For example, broad concepts such as "audience stance" or "audience relevance" in argumentation can be refined into measurable psychological constructs such as personal relevance and mood. These can be operationalized experimentally, offering clearer insights into how audiences respond to different argument types.

Additionally, argumentation theory informs psychology by refining terms used in studies where distinctions between "weak" and "emotional" arguments have often been imprecise.

6.3 Model of Interactional Pathos in Argumentation for Experiments on Argument Strength

In previous sections, I presented how methods from experimental psychology can be applied to answer questions from argumentation theory and rhetoric. In this section,

I would like to present my own study (Konat et al. 2024) analyzing the influence of personal involvement, mood, and emotional appeals on perceived argument strength (PAS).

Let us go back to the Model of Interactional Pathos in Argumentation (MIPA) presented in Part I of this book. Just to reiterate, the model assumes that pathos occurs in persuasive interaction, where the speaker is performing emotional appeal with the intention of persuading the listener. In different parts of this book I propose different methods of operationalization of elements of this model, allowing for the study of pathos with various methodologies. For the study presented in this section, I propose to start with the speaker side of the diagram, who in this experiment would be represented by an experimenter producing a textual communication, therefore emotional appeal will be in a form of a natural language argumentation with three forms of appeal—positive, negative, or neutral. In terms of classical rhetoric, the first two will correspond to pathos, and the neutral to logos. On the side of the hearer we have previously seen a multitude of possibilities—from questionnaires, tests, to EEG and face expressions (and in Part III of this book we will show how to capture those with language analysis). For this study we will capture cognitive effects and emotion via questionnaires: Perceived Argument Strength (Zhao et al. 2011), and questions about personal involvement and mood.

Figure 6.2 shows how MIPA can be adapted to illustrate the experimental setting. On the speaker's side, a stylized scholar figure represents a professor in a classroom who provides an emotional appeal and an argument to students through indirect written materials. The audience is depicted as a group of students, reflecting the actual classroom environment used in the experiment. On the cognitive level, the structure remains the same: the speaker performs an emotional appeal, and the audience experiences an emotional reaction. On the language level, the speaker's message includes positive, negative, and neutral arguments. On the audience side, we measure perceived argument strength (PAS), mood change, and personal involvement.

A useful theoretical tool for this study comes from the concept of personal involvement in the Elaboration Likelihood Model (ELM) (Petty and Cacioppo

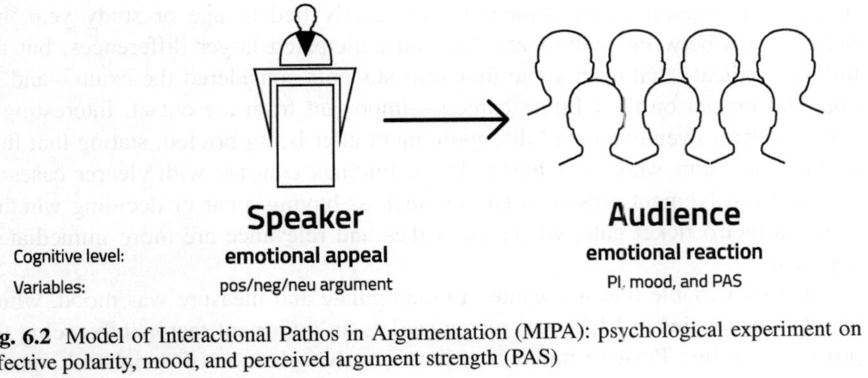

Fig. 6.2 Model of Interactional Pathos in Argumentation (MIPA): psychological experiment on affective polarity, mood, and perceived argument strength (PAS)

1986), a framework frequently used for persuasion studies in psychology. ELM is a widely used theoretical framework in psychology, with numerous studies testing how likely it is that a given argument will be elaborated, i.e. processed via what is referred to as the central route. The central processing route, involving high elaboration according to ELM, occurs when we are about to make a decision on something important to us, something we feel strongly involved in. For example, deciding which ticket gate to use at a metro station is probably not an important decision; we are unlikely to elaborate extensively on which one to choose (and will use the peripheral processing route). However, a decision about which car to buy carries a much heavier load in terms of financial expense, safety, future costs of fuel, etc., so we are less likely to make it impulsively (and will use the central processing route). According to ELM, weak arguments (which are defined as using pathos or ethos as strategies) can be perceived as stronger with low involvement, whereas strong arguments (defined as using logos: facts, evidence, numbers) would appeal more to participants with higher involvement. In our study, we followed these concepts but enriched them with several more layers for better precision and to align them with the needs of argumentation theory.

In our study, we used an experimental setup similar to that employed by Petty et al. (1981). During regular class hours, one of our experimenters entered a classroom and asked psychology students (N = 395) their opinions on the persuasiveness of the arguments supporting the introduction of a new practical final exam. The students were led to believe that this new exam, mandatory before graduation, would be implemented. At the end of the experiment, the students were debriefed, both verbally and in writing, and informed that no such exam was actually planned. This scenario was particularly relevant to the first manipulation we introduced: approximately half of the participants were first-year students, meaning they had four more years before the supposed exam, while the other half were fourth-year students, with just over a year left to prepare. We hypothesized that fourth-year students would exhibit higher levels of personal involvement, while first-year students would show lower levels due to the distant timeline. We measured personal involvement on a 5-point scale (from "not at all important" to "very important"). Surprisingly, the mean scores were 3.1 for the first-year students and 3.6 for the fourth-year students, both with a standard deviation of 1.2. These results suggest that personal involvement is not strictly tied to age or study year, but rather varies between individuals. We had anticipated larger differences, but the findings indicate that even some first-year students considered the exam—and its potential impact on their future careers—important from the outset. Interestingly, some students even expressed disappointment after being briefed, stating that they wished the exam were introduced. These findings contrast with clearer cases of personal involvement presented earlier, such as buying a car or deciding whether to use a metro ticket gate, where the stakes and relevance are more immediately apparent.

Another variable that we wanted to manipulate and measure was mood, which in previous studies in ELM was recognized as an important factor influencing the processing route: Positive mood contributes to the selection of peripheral signals, resulting in less elaboration and more heuristic processing (Hullett 2005; Sinclair et al. 2010). In our study, half of the participants, randomly selected, received a voucher of about 10 EUR, whereas other half received no incentive. Contrary to our expectations, the participants who received the voucher reported no significant differences in mood compared to those who did not receive it. This suggests that mood is not easily influenced by a small incentive in this context, and in an in-class experiment, it may be shaped more by external factors beyond the researcher's control. At the same time, as we will see later, positive emotional appeal in written texts presented to participants consistently led to higher moods, showing that influence with verbal means is possible and can have strong effects.

Our final and most critical manipulation involved the concepts of pathos and logos. Traditionally, within the Elaboration Likelihood Model (ELM), arguments are classified into two categories: neutral, logos-based arguments (considered strong), and those based on emotions or authority, i.e., ethos or pathos (considered weak). For our study, we opt for a more detailed classification. First, we excluded ethos-based arguments as we believe that they require separate investigation. This left us with a distinction between logos-based (neutral) arguments and pathos-based (emotional) arguments. We further subdivided pathos arguments into those with positive and negative emotional appeals. As a result, we created three types of argumentation stimuli that were randomly assigned to the participants. Importantly, each participant was unaware that the person next to them might receive a persuasive text with a different emotional tone. All three argument types followed the structure of practical reasoning, each concluding in a call to action—the necessity of introducing the new final exam. The neutral argument relied on facts and figures, including statements such as: "The study revealed that implementing such an exam increased graduates' self-assessed knowledge by 31%. In comparable institutions without the exam, self-assessed knowledge increased by only 8% over the same period". The positive argument focused on positive language and envisioned a bright future: "We want our students to feel motivated to improve their skills and view the feedback they receive as valuable and conducive to personal development." The negative argument invoked past problems and highlighted fears about an uncertain job market, with statements such as: "The poorly developed careers of graduates also contributed to the negative assessment [of psychology]". To validate our intuition regarding the sentiment of each message, my team used sentiment analysis models which confirmed that each argument conveyed the intended sentiment. I believe this approach enriches the ELM framework by introducing more precise distinctions between ethos and pathos, and also contributes to discussions in argumentation theory by operationalizing concepts from classical rhetoric.

On the concept of argument strength, my team aimed at contributing to better understanding of the empirical approach, as described in Zenker et al. (2020, 2023). The empirical study of argument strength evaluates how audiences perceive arguments, using experimental methods to track cognitive and attitudinal reactions. This approach, unlike formal or theoretical, assesses the effectiveness of the argument based on the responses of the real participants, quantifying perceptions of persuasiveness, credibility, or strength. In the part devoted to empirical approach to argument strength, Zenker et al. (2020) investigates how different arguments about the fear of death are perceived in terms of strength. In their study, Polish university students, proficient in English, evaluated three arguments: the "no subject of harm" argument (asserting that death is not harmful since one cannot experience pain after death), the "symmetry" argument (suggesting no fear of death by analogy with the time before birth) and the "squandering" argument (critiquing a lack of value placed on life). The participants rated each argument on dimensions such as believability, confidence, and importance. The findings indicated that the "no subject of harm" and "symmetry" arguments were perceived as somewhat stronger than the "squandering" argument, supporting the idea that the latter may function more effectively as a dialectical response than as a standalone argument. A tool used in Zenker et al. (2020), and also in the study presented here is the Perceived Argument Strength (PAS) questionnaire (Zhao et al. 2011), which includes indicators such as believability, importance, and agreement, enabling researchers to evaluate argument strength in fields ranging from public health to education. Its adaptability makes it valuable for studies that need direct participant feedback on arguments' impact.

Going back to the study of my team (Konat et al. 2024) our results reveal first that personal involvement of the audience is crucial in how they assess argument strength—participants self-reporting higher personal involvement (i.e. those who marked "the topic is very important to me") assessed all arguments as stronger than those participants who reported lower levels of personal involvement. Secondly, mood is another important factor—participants who self-report higher mood in pretest (i.e. before the start of experiment) assessed all arguments as stronger than those participants who reported lower mood. Last but certainly not least, in the context of personal involvement and mood we were able to observe how different types of emotional appeal (positive and negative, vs. no appeal—neutral) influence perceived argument strength. Our results show that even though personal involvement is the strongest predictor, still argumentation with positive emotional appeal was deemed stronger than that with neutral valence, and negative emotional appeals were judged as weakest. The difference was particularly great among the participants who reported high levels of personal involvement.

Going back to Walton's (1992) approach to the strength of emotional argument mentioned earlier, do emotional appeals make arguments seem stronger? Certainly positive emotional appeals might make arguments appear stronger; however, negative ones might make them appear weaker, even weaker than neutral arguments. Our study also provided more empirical evidence towards the crucial role of audience relevance, and the role of pathos in it, as presented by Tindale (2015), who long argued for the need of understanding of audience's emotions, biases, and values for effective rhetorical argumentation. As our study shows, the two audience-related factors (personal involvement and mood) are not easy to predict or manipulate—we tried to predict personal involvement by assuming that senior year students will be more involved, but it turned out it is a more personalized question, and some first year persons can be more involved; similarly we tried to manipulate the mood with a voucher, but mood also seems to be influenced by multitude of different factors outside of speakers control. This shows that "knowing your audience" is becoming more important in this context.

Another issue lies in what Walton called "dark-side commitments" (Walton 1992, p. 255), or existing views and predispositions of the audience. In appealing to the mind of the audience, the speaker, the rhetor, is confronted with pre-existing stances, views, experiences, and emotions. All those to whom they do not have access. Every emotional appeal performed for rhetorical gain "lands" on this existing background or predisposition, an issue of which both Aristotle (when writing about pathos) and Walton (when writing about emotional appeals) were aware. Transforming this observation in our experimental work, we attempted to control for two elements of these predispositions: mood and personal involvement in the issue at stake. However, we are aware that this background is much richer and that individual differences, both stable like personality traits and temporary like daily mood, can influence strongly the effectiveness of emotional appeals.

This empirical approach and the PAS scale have limitations, as mentioned in Zenker et al. (2020). Experimental evaluations may miss context-specific features of argumentation that dialectical approaches emphasize. For example, complex arguments with dialectical interactions or cultural dependencies may need adjustments to fit the PAS model, and some indicators (such as novelty) may not suit every argument type.

To sum up, in our experimental study we wanted to provide empirical evidence to answer the question whether emotional appeals may make arguments appear stronger than they really are. Our results suggest that positive appeal can make an argument look stronger, but crucial factors are audience-related: personal involvement and mood. We also show, that next to normative, dialectical, and computational methods of assessing argument strength, an empirical method with experiments involving human participants can be a useful tool.

One limitation of the experimental approach is its weak connection to real-world contexts. The controlled nature of experimental stimuli can make them appear artificial, and participants are often confined to answering questionnaires rather than engaging in real discussions. While this provides rigour, it may miss the complexity of genuine argumentative interactions. Spontaneous conversations in natural language corpora, despite their own limitations, offer a more realistic perspective. In the subsequent Chaps. 8 and 9, I will demonstrate how these approaches can complement experimental methods and provide a fuller understanding of emotional appeals in argumentation.

In summary, do emotional arguments appear stronger than they are? Yes, they might: including positive emotional appeal in the weak argument will increase its perceived strength. However main factors contributing to the perceived strength of an argument are related to the audience and not argument itself. First, for an audience to which the topic is important—every argument will appear as strong. Second, for persons in better mood, also all arguments will appear stronger. However these audience factors are not easy to manipulate or change—in our study we failed to manipulate them by providing vouchers to selected participants or by selecting specific year of study. It does look like probably they are more idiosyncratic and individual. This leads us back to Aristotle's concept of pathos as appealing to the existing emotions to the audience—when performing emotional appeals it is more important to know the audience and their priorities and moods, than to try to elicit emotions particularly useful to the speaker. This result contributes to our understanding of audience effects, in particular the idea of audience relevance by Tindale (2015)—the persuasiveness of an argument does depend mainly on audience engagement. This in turn leads us to an empirical understanding of the strength of the argument—this study contributes through the use of PAS (Konat et al. 2024). This all brings the audience effects to the forefront of the strength of argument.

6.4 Model of Interactional Pathos in Argumentation for Experimental Competitive Debating

This section presents an empirical extension of the Model of Interactional Pathos in Argumentation (MIPA) to the context of competitive student debating. Building on the core premise that pathos is a dynamic, intersubjective process between speaker and audience, we investigate how emotional appeals are constructed and evaluated within structured rhetorical exchanges.

Figure 6.3 shows an adaptation of the interactional pathos model to an experimental setting involving competitive student debating. As in earlier diagrams, the Cognitive level remains unchanged: the speaker formulates an emotional appeal, and the audience experiences an emotional reaction. This illustrates how the same underlying process of pathos can take on different surface-level forms depending on the communicative context.

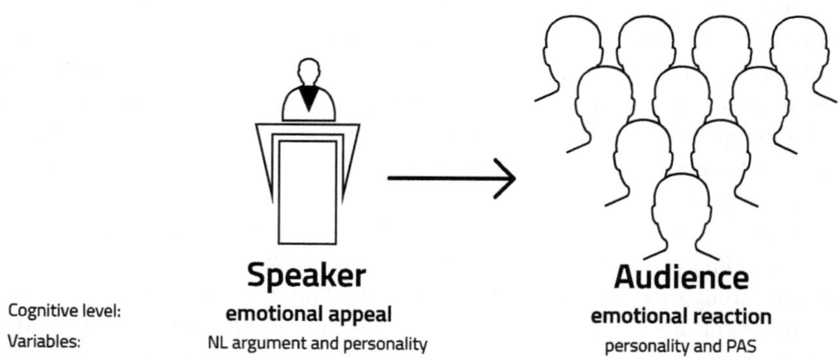

Fig. 6.3 Model of Interactional Pathos in Argumentation (MIPA): speaker variables—argument style, personality traits, and PAS

In this case, the speaker is a student taking part in a structured debate, addressing their peers. The speaker uses arguments with emotional valence (appealing to categorical emotions like fear, joy, or disgust) to influence the audience. On the Language level, this is represented through the NL (natural language) argument. The diagram indicates that in this natural experiment, we observed and measured variables for both speakers and hearers. The audience consists of fellow students acting as listeners and evaluators. Their reactions are measured through perceived argument strength (PAS) and personality match, assessing how closely their personality traits align with the speaker's.

This chapter is based on a study from my team (Obrębska et al. 2025), where emotional appeal (as formulated by the speaker) and emotional reactions (as experienced by the audience) are investigated with variables central to the MIPA framework: argument style, personality traits, and perceived argument strength.

This study employs a natural experiment design, which refers to an empirical investigation conducted in a real world setting where the researcher has limited control over variables, allowing participants to engage in spontaneous behavior. Unlike the previous experiments presented here, natural experiments preserve the social and contextual complexity of the phenomena under study. Here, persuasive exchanges unfolded during a structured debate between students who assumed the roles of both speaker and hearer. By observing and measuring speaker- and hearer-side variables in this minimally manipulated environment, the study captures interactional pathos in its authentic, emergent form.

With this specific setting, our aim was to examine whether speaker traits, specifically argumentation style and personality, affect perceived argument strength, and whether psychological similarity between speaker and listener modulates this effect. The experiment was conducted during a university summer school on public speaking, organized specifically for this purpose. Seventeen psychology and cognitive science students participated in British Parliamentary debates after two days of intensive rhetorical training. All participants completed the NEO-Five Factor Inventory - NEO-FFI personality inventory (Costa and McCrae 1992) and were recorded during the final debate rounds, with Perceived Argument Strength (PAS, Zhao et al. 2011) scores collected after each speech from audience-judges.

The emotional argumentation style was operationalized through sentiment analysis and hierarchical clustering of valence trajectories in speeches. Two recurrent styles appeared: one emotionally intense and negatively marked, especially at the opening; the other transitioning between valences with a tendency toward neutral tone. These stylistic profiles were treated as speaker-level variables, alongside their Big Five personality dimensions.

The study formulated five hypotheses aimed at exploring the interactional dynamics of persuasive discourse. First, it hypothesized that speaker personality traits would predict the argumentation style they employ (H1), and that these traits would also influence how strong their arguments are perceived to be (H2). It further proposed that the style of argumentation itself might directly affect perceived argument strength (H3). Finally, the study examined whether psychological similarity between speaker and listener, either in argumentation style (H4) or in personality traits (H5), would enhance the perceived persuasiveness of the arguments. These hypotheses reflect the core tenets of the MIPA framework, emphasizing how both individual dispositions and interpersonal alignments contribute to pathos in argumentation.

None of the hypotheses were statistically supported. Neither speaker personality traits nor argumentation style significantly predicted PAS. Psychological similarity, whether in style or personality, did not consistently enhance perceived persuasive strength. While some individual listeners rated similar speakers more favorably, others showed the opposite pattern or no pattern at all. These results suggest that the effects of psychological matching are not uniformly positive, but heterogeneous, possibly shaped by factors such as self-construal or context-specific goals. However, the significance of these findings lies not in the confirmation of effects, but in the methodological shift they represent. Unlike previous work on personalized persuasion that relies on pre-written messages and controlled exposure, this study preserves the spontaneity and turn-taking structure of real argumentative interaction, on the cost of no control over variables. Participants took on both speaker and judge roles, allowing for symmetric measurement of traits, styles, and evaluative responses, consistent with the interactional framework of MIPA.

Despite the small sample size and reduced experimental control, the design offers ecological validity. Emotional appeals unfolded naturally, with valence contours reflecting the speaker's strategic use of pathos across debate segments. This contrasts with standard persuasion studies that flatten emotional variation to single cues or static message frames. The observed variation in emotional strategies, especially in negative vs. positive framing, points to the need for models of argumentation to account for affective dynamics over time.

Our data suggest that pathos, as captured in MIPA, is deeply shaped by speaker-hearer interaction but resists simple mappings based on dispositional similarity. While psychological matching remains a compelling theoretical construct, its operationalization in real discourse may require more variables, including affective states, argument content, or self-relevance, as well as larger samples to detect interaction effects. The implications for the computational modeling of argumentation are discussed in Chap. 9, where we analyze the transcripts through transformer-based models and further refine the operationalization of argumentation style. The current chapter lays the conceptual groundwork by emphasizing the importance of emotion, interaction, and ecological validity in the study of argument strength.

In summary, this chapter has shown how emotional appeals in argumentation can be studied empirically through lexicon-based analyses and experimental methods grounded in psychological theory. By integrating dimensional and categorical models of emotion with the interactional framework of pathos, it becomes possible to trace how emotion-eliciting words function rhetorically across languages and contexts. Experimental results further show that emotional arguments, particularly those with positive valence, can enhance perceived argument strength, although this effect is consistently modulated by audience-related factors such as mood and personal involvement. These findings support a pragmatically grounded and empirically testable model of pathos, while also pointing to the complexity of audience effects and the limits of experimental control. Together, the lexicon and experimental approaches presented here form a complementary empirical toolkit for studying emotional appeals, providing a basis for future work in both corpus analysis and computational modeling, as explored in the following chapters.

As we close Part II, several conclusions are worth reiterating. First, as multiple studies claim, language provides a crucial window into conceptual and affective processes. Whether through emotion terms, metaphor, or emotionally marked lexical choices, discourse reveals how speakers encode and elicit emotional responses. Second, affective lexicons demonstrate that words can function as emotional stimuli, shaping audience reactions in predictable ways. Third, the Model of Interactional Pathos in Argumentation (MIPA) offers a flexible framework for studying these effects across contexts, from controlled experiments to naturalistic debates. Finally, empirical findings show the importance of audience-related factors, such as mood, involvement, and personality, as emotional appeals do not operate in a vacuum, but are embedded in interaction. These conclusions provide the conceptual and methodological foundation for Part III, where we turn to corpus analysis and computational modeling to examine how emotional appeals function in real-world discourse.

References

Benlamine MS, Villata S, Ghali R, Frasson C, Gandon F, Cabrio E (2017) Persuasive Argumentation and Emotions: An Empirical Evaluation with Users. Lecture Notes in Computer Science (including subseries Lecture Notes in Artificial Intelligence and Lecture Notes in Bioinformatics) 10271:659–671, https://doi.org/10.1007/978-3-319-58071-5_50

Benlamine S, Chaouachi M, Villata S, Cabrio E, Frasson C, Gandon F (2015) Emotions in argumentation: an empirical evaluation. In: Twenty-Fourth International Joint Conference on Artificial Intelligence

Blanchette I (2006) The effect of emotion on interpretation and logic in a conditional reasoning task. Memory & Cognition 34(5):1112–1125, https://doi.org/10.3758/BF03193257

Bradley MM, Lang PJ (1999) Affective norms for English words (ANEW): Instruction manual and affective ratings. Tech. Rep. Technical Report C-1, University of Florida, Center for Research in Psychophysiology

Cerutti F, Tintarev N, Oren N (2014) Formal arguments, preferences, and natural language interfaces to humans: an empirical evaluation. In: ECAI 2014: Frontiers in Artificial Intelligence and Applications, vol 263, pp 207–212, https://doi.org/10.3233/978-1-61499-419-0-207

Cigada S (2019) Emotions in argumentative narration: The case of the Charlie Hebdo attack. Informal Logic 39(4):401–431

Costa PT, McCrae RR (1992) The five-factor model of personality and its relevance to personality disorders. Journal of Personality Disorders 6(4):343–359, https://doi.org/10.1521/pedi.1992.6.4.343

Dung PM (1995) On the acceptability of arguments and its fundamental role in nonmonotonic reasoning, logic programming and n-person games. Artificial intelligence 77(2):321–357

van Eemeren FH, Garssen B, Meuffels B (2009) Fallacies and Judgments of Reasonableness: Empirical Research Concerning the Pragma-Dialectical Discussion Rules, Argumentation Library, vol 16. Springer

Ervas F (2019) Metaphor, ignorance and the sentiment of (ir)rationality. Synthese https://doi.org/10.1007/s11229-019-02489-y

Ervas F (2021) The double framing effect of emotive metaphors in argumentation. Frontiers in Psychology 12:628460, https://doi.org/10.3389/fpsyg.2021.628460

Gajewska E, Budzynska K, Konat B, Koszowy M, Kiljan K, Uberna M, Zhang H (2024) Ethos and pathos in online group discussions: Corpora for polarisation issues in social media. arXiv preprint arXiv:240404889 2404.04889

Guasch M, Ferré P, Fraga I (2016) Spanish norms for affective and lexico-semantic variables for 1,400 words. Behavior Research Methods 48:1358–1369, https://doi.org/10.3758/s13428-015-0684-y, https://link.springer.com/article/10.3758/s13428-015-0684-y, published online: 05 November 2015

Heit E, Rotello CM (2012) The pervasive effects of argument length on inductive reasoning. Thinking and Reasoning 18(3):244–277, https://doi.org/10.1080/13546783.2012.695161

Hornikx J, Hahn U (2012) Reasoning and argumentation: Towards an integrated psychology of argumentation. Thinking & Reasoning http://www.tandfonline.com/loi/ptar20, version of record first published: 11 June 2012

Hullett CR (2005) The impact of mood on persuasion: A meta-analysis. Communication Research 32(4):423–442, https://doi.org/10.1177/0093650205277317

Imbir KK (2016) Affective norms for 4900 Polish words reload (ANPW_R): Assessments for valence, arousal, dominance, origin, significance, concreteness, imageability and, age of acquisition. Frontiers in psychology 7:1081

Konat B, Budzynska K, Saint-Dizier P (2016) Rephrase in argument structure. In: Saint-Dizier P, Stede M (eds) Foundations of the Language of Argumentation. COMMA 2016 Workshop, University of Potsdam, pp 32–39

Konat B, Obrębska M, Gajewska E, Dembska N (2024) Understanding perceived argument strength: The effects of personal involvement, mood, and pathos appeals. In reviews. Submitted Manuscript

Koszowy M, Oswald S, Budzynska K, Konat B, Gygax P (2022) A pragmatic account of rephrase in argumentation: linguistic and cognitive evidence. Informal Logic 42(1):49–82, https://doi.org/10.22329/il.v42i1.7212

Lang PJ, Bradley MM, Cuthbert BN (2005) International affective picture system (iaps) [database record]. https://doi.org/10.1037/t66667-000, APA PsycTests

Macagno F, Walton D (2014) Emotive Language in Argumentation. Cambridge University Press, https://doi.org/10.1017/CBO9781139565776

Mohammad S, Turney P (2010) Emotions evoked by common words and phrases: Using mechanical Turk to create an emotion lexicon. In: Proceedings of the NAACL HLT 2010 workshop on computational approaches to analysis and generation of emotion in text, pp 26–34

Mohammad SM, Turney PD (2013) Crowdsourcing a word–emotion association lexicon. Computational intelligence 29(3):436–465

Monnier C, Syssau A (2014) Affective norms for French words (FAN). Behavior Research Methods 46(4):1128–1137, https://doi.org/10.3758/s13428-013-0431-1

Montefinese M, Ambrosini E, Fairfield B, Mammarella N (2014) The adaptation of the affective norms for English words (ANEW) for Italian. Behavior Research Methods 46:887–903, https://doi.org/10.3758/s13428-013-0405-3, https://link.springer.com/article/10.3758/s13428-013-0405-3, published online: 23 October 2013

Obrębska, M and Konat, B and Kleka and Kiljan, K and Dembska, N, Personality traits and argumentation style as factors in perceived argument strength in natural conditions of competitive debating. The role of psychological matching of speaker and listener. Available at SSRN: https://ssrn.com/abstract=5416998 or http://dx.doi.org/10.2139/ssrn.5416998

Osgood CE, Suci GJ, Tannenbaum PH (1957) The Measurement of Meaning. University of Illinois Press, Urbana, IL

Petty RE, Cacioppo JT (1986) The Elaboration Likelihood Model of Persuasion. Advances in Experimental Social Psychology 19:123–205, https://doi.org/10.1016/S0065-2601(08)60214-2

Petty RE, Cacioppo JT, Goldman R (1981) Personal involvement as a determinant of argument-based persuasion. Journal of Personality and Social Psychology 41(5):847–855, https://doi.org/10.1037/0022-3514.41.5.847

Plantin C (2019) Tense arguments: questions, exclamations, emotions. Informal Logic 39(4):347–371

Polberg S, Hunter A (2018) Empirical evaluation of abstract argumentation: Supporting the need for bipolar and probabilistic approaches. International Journal of Approximate Reasoning 93:487–543, https://doi.org/10.1016/j.ijar.2017.11.008

Rahwan I, Madakkatel MI, Bonnefon JF, Awan RN, Abdallah S (2010) Behavioral experiments for assessing the abstract argumentation semantics of reinstatement. Cognitive Science https://doi.org/10.1111/j.1551-6709.2010.01123.x, first published: 03 November 2010

Sarli L, Justel N (2022) Emotional words in Spanish: Adaptation and cross-cultural differences for the affective norms for English words (anew) on a sample of Argentinian adults. Behavior Research Methods 54:1595–1610, https://doi.org/10.3758/s13428-021-01666-5, https://link.springer.com/article/10.3758/s13428-021-01666-5, published online: 10 September 2021

Schumann J, Zufferey S, Oswald S (2019) What makes a straw man acceptable? three experiments assessing linguistic factors. Journal of Pragmatics 141:1–15, https://doi.org/10.1016/j.pragma.2018.12.008

Sinclair RC, Moore SE, Mark MM, Soldat AS, Lavis CA (2010) Incidental moods, source likeability, and persuasion: Liking motivates message elaboration in happy people. Cognition and Emotion 24(6):940–961, https://doi.org/10.1080/02699930903000206

Soares AP, Comesaña M, Pinheiro AP, Simões A, Frade CS (2012) The adaptation of the affective norms for English words (anew) for European Portuguese. Behavior Research Methods 44:256–269, https://doi.org/10.3758/s13428-011-0131-7, https://link.springer.com/article/10.3758/s13428-011-0131-7, published online: 13 July 2011

Stevenson RA, Mikels JA, James TW (2007) Characterization of the Affective Norms for English Words by discrete emotional categories. Behavior Research Methods 39(4):1020–1024, https://doi.org/10.3758/BF03192999

Tindale CW (2015) The Philosophy of Argument and Audience Reception. Cambridge University Press, Cambridge

Villata S, Cabrio E, Jraidi I, Benlamine S, Chaouachi M, Frasson C, Gandon F (2017) Emotions and personality traits in argumentation: An empirical evaluation. Argument & Computation 8(1):61–87

Walkowiak MT, Imbir KK (2018) The role of the origin (automatic vs. reflective) of affective state for the effectiveness of persuasion based on strong vs. weak arguments. Roczniki Psychologiczne/Annals of Psychology 21(1):9–33, https://doi.org/10.18290/rpsych.2018.21.1-2

Walton D (1992) The Place of Emotion in Argument. The Pennsylvania State University Press, University Park, PA

Warriner AB, Kuperman V, Brysbaert M (2013) Norms of valence, arousal, and dominance for 13,915 English lemmas. Behavior Research Methods 45(4):1191–1207

Wierzba M, Riegel M, Kocoń J, Miłkowski P, Janz A, Klessa K, Juszczyk K, Konat B, Grimling D, Piasecki M, et al. (2021) Emotion norms for 6000 Polish word meanings with a direct mapping to the Polish wordnet. Behavior Research Methods pp 1–16

Younis R, Fernandes DdO, Gygax P, Koszowy M, Oswald S (2023) Rephrasing is not arguing, but it is still persuasive: An experimental approach to perlocutionary effects of rephrase. Journal of Pragmatics 210:12–23, https://doi.org/10.1016/j.pragma.2023.03.010

Zenker F, Debowska-Kozlowska K, Godden D, Selinger M, Wells S (2020) Five approaches to argument strength: probabilistic, dialectical, structural, empirical, and computational. In: Proceedings of the 3rd European Conference on Argumentation, pp 653–674

Zenker F, Dębowska-Kozłowska K, Godden D, Selinger M, Wells S (2023) Thou shalt not squander life - comparing five approaches to argument strength. Studies in Logic, Grammar and Rhetoric 68(81), https://doi.org/10.2478/slgr-2023-0007, http://creativecommons.org/licenses/by/4.0

Zhao X, Strasser A, Cappella JN, Lerman C, Fishbein M (2011) A measure of perceived argument strength: Reliability and validity. Communication Methods and Measures 5(1):48–75

Part III
Artificial Intelligence for Pathos

Chapter 7
AI Approaches to Argumentation

Abstract This chapter introduces computational approaches to natural language argumentation, serving as a foundation for the pathos mining methods presented in subsequent chapters. It outlines the two main strands of computational argumentation: formal argumentation frameworks and natural language argument mining. The chapter focuses on the latter, reviewing tools, corpora, and techniques used to model and extract argumentative structures from texts. It describes the progression from lexical methods through supervised machine learning to transformer-based and large language models (LLMs), showing how each has contributed to automating argument mining. The chapter presents recent developments in the use of LLMs for classification tasks, annotation support, and persuasive discourse analysis, while critically evaluating claims about intentionality and strategy in AI-generated argumentation. It argues that although LLMs can emulate human-like argumentative behavior, their outputs remain grounded in statistical pattern matching. Finally, it identifies how LLMs and other computational tools can be used to model emotional appeals (pathos) in argumentation, in preparation for the pathos mining framework developed in later chapters.

Keywords Argumentation mining · Natural language processing · Machine learning models · Transformer models · Large language models

7.1 Lexicon-Based Methods in Argumentation Mining

Computational approaches to argumentation have evolved rapidly, moving from lexical heuristics to machine learning and large language models (LLMs). This chapter introduces key methods in natural language argument mining, presenting how these technologies identify and simulate argumentative structures. It also examines their limitations, particularly in attributing intentionality or strategy to AI outputs.

Argumentation mining is one of the key elements of computational approaches to natural language argumentation and one that has an established place within the argumentation community. As a survey defines it: "Argument mining is the

B. Konat, *Emotional Appeals in Argumentation*, Argumentation Library 47,
https://doi.org/10.1007/978-3-032-06332-8_7

automatic identification and extraction of the structure of inference and reasoning expressed as arguments presented in natural language" (Lawrence and Reed 2019, p. 765). From its modest beginnings, the area has developed to include a comprehensive survey (Lawrence and Reed 2019) and a series of workshops (11 at the time of writing) held in conjunction with prestigious conferences. Technologies applied in argument mining will therefore serve as the background and introduction to my proposal for pathos mining, as they represent analogous techniques for extracting complex persuasive intentions expressed in natural language.

The lexical method is a well-established approach in argumentation mining, originating from a time when the computational power available today was not accessible. Still, I will present reasons why I find it useful in the present context. An example of how the lexical method can be applied to this task is provided in a paper by Lawrence and Reed (2015). In this study, the authors demonstrate how combining lexical indicators of argumentation with simple Natural Language Processing (NLP) techniques can achieve high accuracy in automated identification of argumentative structures in natural language. They use lists of words that indicate intentional argumentation by the speaker, such as "because", "therefore", "after", "for", "since", "when", "assuming", and "so", among others. Using this approach, they achieve very high precision, meaning that nearly all instances identified as arguments are indeed arguments. However, the method has a significant drawback: it has very low recall. Low recall means that if the lexical method identifies a sentence as an argument (e.g., based on the presence of the word "because"), we can be sure that it is an argument. However, when the method omits a sentence, we cannot be sure whether the sentence is not an argument or simply was not identified by the method. This means that only a small fraction of all arguments in the corpus is identified, leaving many undetected. To address this limitation, Lawrence and Reed (2015) supplemented their approach with an additional method (topic identification), combining the two for improved results.

The main conclusion from this study, as it relates to the pathos mining methods discussed here, is that lexical methods deliver high precision but low recall. This means that, similar to the findings of Lawrence and Reed (2015), when a result of the pathos mining method identifies an emotion-eliciting word within an argument, we can be confident of its presence and its role in eliciting emotions in the audience. However, as with argumentation mining, we cannot determine whether the method has omitted other cases that lack words from the employed lexicon. This dual nature, its strengths in precision and limitations in recall, defines the utility and constraints of lexicon-based methods in argumentation studies. There is also another takeaway from this case. Just like lexical methods, both classic language models based on machine learning (such as the sentiment analysis models discussed in the next chapter) and Large Language Models (LLMs, such as GPT, discussed later in this Part III) operate on the linguistic surface. They analyze words, sentences, and entire texts, informed by the context derived from their training process. Unlike lexical methods, which rely on raw words, these models process mathematical representations of relationships between words and phrases.

This approach makes language models more effective in identifying arguments even in the absence of explicit connectives such as "because". For example, if new examples lack the word "because" but are sufficiently similar to training examples where the word appeared and was labeled as arguments, the model can still recognize them as such. In this sense, language models, and in particular LLMs, are described as intelligent, though their "knowledge" is entirely derived from the extensive textual data used during training, and should not be misunderstood for conscious or human-like intelligence. This is particularly important for pathos mining, as I will emphasize that language models can correctly identify the appeal to emotions in text, just like human would do, however, they do not feel or understand these emotions, instead just emulating human verbal behaviors.This continuity, from lexical methods to classical models to LLMs, shapes their application to both argumentation mining and the emerging field of pathos mining, which will be explored in later chapters. What constitutes this continuity is the shared focus on patterns in language, albeit at increasingly complex levels of abstraction.

7.2 Machine Learning Models for Argument Structure

For argumentation mining, the continuous development of various machine learning (ML) models was the source of the thriving of this discipline. ML models have been crucial in the evolution of argument mining, enabling the automated extraction of argumentative components and their relationships from natural language texts and overcoming limitations of lexical methods described earlier. These models provide adaptability across domains and tasks, capturing patterns in language beyond surface-level features, and this is why they were gradually replacing lexical methods.

Supervised learning models, such as Support Vector Machines (SVMs), Decision Trees, and Naïve Bayes, formed the foundation of early argument mining. These models require annotated corpora for training, where features such as linguistic markers, dependency parses, and discourse signals are designed manually to inform classification tasks (Levy et al. 2014). Although these traditional approaches are limited in scalability due to their reliance on high-quality annotated data, they remain effective for tasks with clearly defined linguistic features. The development in neural networks has significantly improved the accuracy of argument mining tasks. Convolutional Neural Networks (CNNs) and Recurrent Neural Networks (RNNs), including Long Short-Term Memory (LSTM) networks, have been applied to tasks such as detecting argument components and predicting relation classification (Stab and Gurevych 2017). These architectures are great at capturing both local and sequential dependencies within text, offering improvements in modeling complex argumentative structures compared to earlier models.

The revolution came with the introduction of transformer models, such as BERT (Bidirectional Encoder Representations from Transformers). Fine-tuned BERT models have been used in tasks such as stance detection and argumentative discourse parsing, achieving state-of-the-art results by employing contextual embeddings, enabling the inclusion of argumentative context in a manner that more closely mirrors human reasoning (Choudhury et al. 2021). These models were also featured in the 2024 Argument Mining Workshop (ArgMin). For example, shared tasks such as DialAM-2024 demonstrated how models incorporating dialogue-specific context can reconstruct argumentative structures from natural conversations (Ruiz-Dolz et al. 2024). Semi-supervised learning, using partially labeled datasets, is also increasingly popular for tasks where annotated data is scarce. At ArgMin 2024, the PerspectiveArg shared task illustrated how semi-supervised approaches could be used to retrieve arguments aligned with specific socio-cultural perspectives, highlighting the growing interest in perspective-aware argument retrieval (Falk et al. 2024).

The presence of these shared task at the most recent Argument Mining workshop shows how this discipline has matured, due to the development of both theoretical advances (what constitutes an argument, including its dialogical and perspective contexts), but also technical development of machine learning models used for this task. For pathos mining, we do not yet have such developments, as the area is just emerging. There are no developed and trained machine learning models that will automatically analyze emotional appeals, and there is not enough highly annotated corpora to effectively train one. What is available, however, are numerous, multilanguage, trained on large dataset models of expressed emotions or sentiment. This is why in our study we will use already trained models for sentiment and emotion recognition to analyze emotions expressed by the audience in reaction to emotional appeals by the speaker in Chap. 8.

7.3 Large Language Models in Argumentation Mining

The progression from traditional machine learning models to large-scale neural networks has been a significant turning point in text analysis and argument mining. While earlier approaches, such as previously mentioned BERT (Bidirectional Encoder Representations from Transformers), introduced transformer-based architectures capable of high quality performance for specific tasks, large language models like GPT (Generative Pre-trained Transformer) represent the next evolutionary step. Both BERT and GPT share underlying technologies, namely neural networks, as well as the Transformer architecture indicated by the "T" in their names. These models are trained on vast amounts of textual data and rely on probabilistic methods to identify patterns. However, the key distinction lies in scale: where BERT operates with approximately 340 million parameters, GPT-4 is suggested to have about 2 trillion parameters.[1] This massive increase in size, computational power, and training data has resulted in emergent capabilities, including the ability to emulate human-like communication and assist in human

[1] At the time of writing, the precise parameter count of GPT-4 remains undisclosed by OpenAI.

annotation tasks. Despite their impressive results, it is still crucial to remember that these systems, commonly labeled as "AI", are—just like previous models—fundamentally probabilistic tools, generating text based on learned patterns rather than genuine reasoning or understanding.

With these caveats, Large Language Models (LLMs) can be highly useful for the emerging field of pathos mining, as they require little data for fine-tuning, and high-quality prompting can be derived from annotation schemes. In this section, we will first examine how LLMs have been used so far for tasks related to argumentation mining and the classification of complex language phenomena. Later, I will propose my own approach for using LLMs for pathos mining.

A significant challenge with LLMs is their non-deterministic nature, as models may produce different outputs on identical tasks across repeated queries. Yet, as with human crowd-sourced annotations, which also involve variability, LLM outputs can be validated through inter-annotator agreement metrics. A popular method is Cohen's Kappa and Fleiss' Kappa to assess consistency between model outputs and human annotations, ensuring reliability even when outputs differ slightly or F1 score to compare automatic with manual annotation (Şahinuç et al. 2024).

Current research covers the use of personalized LLMs for hate speech detection and emotion recognition (Woźniak et al. 2024). For classification task (i.e. manual annotation) they show best results when the prompt included 1 or 2 human annotations (examples). However, in earlier works the same team indicated that, especially for complex task, still specialized models, trained for one classification task, perform better than general LLM (Kocoń et al. 2023). So we always should be mindful of their limitations when using LLMs, focus on high quality of the prompt and good domain knowledge; it is also useful to have at one's disposal a set of high-quality human annotations.

A very interesting approach, which combines emotions, moral dictionary and persuasion, comes from the paper by Carrasco-Farré (2024), where the author analyzes the persuasiveness of texts generated by large language models (LLM). What is interesting in this study, just like in ours, is that it combines theoretical grounding (in classical rhetoric and communication studies) with computational approaches and quantitative analysis, alongside human participants evaluating output. As is the main thesis throughout this book, this combination of methodologies is currently the most useful for studying complex phenomena such as the role of pathos in persuasion.

The relationship between emotions and moral values is purposefully left out of this book, as it constitutes a separate, rich, and complex field. However, recent studies using the Moral Foundations Dictionary (MFD; Haidt and Joseph 2004), a lexicon of words and phrases categorized by moral values based on Moral Foundations Theory, show this intersection. This theory identifies key moral foundations such as care/harm, fairness/cheating, loyalty/betrayal, authority/subversion, and sanctity/degradation. The MFD has been widely used in computational studies to analyze how moral language appears and operates in various forms of argumentation and communication. Studies using MFD are proliferating. A study in "Argumentation" (Landowska et al. 2024) uses MFD for the lexical analysis of moral argumentation. Similarly, the previously mentioned (Carrasco-Farré 2024) uses MFD to analyze what the author calls "human and LLMs" "persuasive strategies." Although the study is interesting and brings new insights, I would like to emphasize the perspective I insist on in this book: language models do not perform persuasion, as they are not sentient beings and do not have what is needed to perform persuasive speech, an intention to persuade a person. LLMs can emulate human verbal behaviors and, by doing so, generate (Carrasco-Farré 2024) persuasive speech or recognize it, just as a human annotator would. This capability is utilized in the analysis that follows this chapter. However, unlike the author in Carrasco-Farré (2024), I argue there is no "strategic use" of argumentation in LLM-produced persuasive speech, as they have no strategy to achieve a goal, as they lack goals other than mimicking human verbal behaviors. In all analyses involving LLMs (or generative AI), we should remain aware of this distinction. The claim that LLMs employ persuasive strategies overlooks their fundamental lack of intentionality. As nonsentient systems, LLMs do not form goals or strategies but only replicate patterns of human language. Although some suggest that LLMs demonstrate intentionality in persuasive contexts, this interpretation misattributes human traits to algorithmic processes, which operate without purpose or awareness. Assigning intentionality to LLMs risks anthropomorphizing these tools, which could mislead analyses by conflating mechanical mimicry with genuine strategic thinking.

van Rooij et al. (2024) provides a clear critique of the current state of AI systems, arguing that they cannot be treated as explanatory models of human cognition. She emphasizes that while these systems may attempt to mimic human behavior, their primary design is task-oriented with practical goals, rather than offering a comprehensive representation of the human mind. This perspective stems from her cautious stance during the recent "AI hype". Supporters of the view that large language models (LLMs) possess intentionality often base their arguments on the models' sophisticated behavior, which can appear deliberate or strategic. However, this interpretation conflates the models' ability to predict and generate contextually appropriate outputs with genuine intentionality or goal-directed behavior. In this context, Chang et al. (2024) present an extensive survey highlighting a wide range of tasks where LLMs have been applied. These applications span traditional domains like natural language processing and fields such as medicine and engineering, often yielding remarkable results.

Large Language Models (LLMs) have also potential applications in classical argument mining tasks. In a recent Argument Mining workshop, Saha and Srihari (2024) presented a study applying the Mistral LLM to extract argument and dialogue reconstructions based on Inference Anchoring Theory from natural language. However, their results remain lower than those achieved by specialized models for argumentation mining. Similarly, Kocoń et al. (2023) observed that while LLMs can perform a wide range of tasks, they often lack domain-specific precision, illustrated in the title "jack of all trades, master of none".

The paper by Bonaldi et al. (2024) explores whether adding constraints (commonly called "guardrails") to language models affects the perceived strength of their hate speech opposing, as compared to creating counter-arguments, similar to those created by humans. The authors focus on how automated systems could moderate online platforms, such as forums or comment sections, when responding to hate speech. For instance, should the system simply label hate speech as unacceptable, or should it attempt to counter the hateful claim with a reasoned response? The study uses a language model Mistral, installed locally to give the researchers more control over its functioning, though it is less powerful than more advanced models like GPT-4. The authors focus on guardrails, which are rules programmed into the system to ensure it avoids generating harmful or inappropriate content. These restrictions are widely used to address ethical concerns about language models creating problematic outputs. Interestingly, the study finds that users perceive counter-arguments as more persuasive and useful than generic responses like "This language is inappropriate and will not be accepted". The authors suggest that advances in automated argument generation could improve moderation systems, making them more effective in fostering constructive dialogue.

One noteworthy aspect of this paper is the authors' use of GPT model to automatically identify premises and conclusions in arguments as a first step of their annotation process. While this is an innovative choice, it is surprising given that the automatic identification of argumentative structures remains a major challenge in computational argumentation. For further insights into how language models like GPT are being applied in this field, see the earlier discussion on mining arguments with LLMs. Returning to Bonaldi et al. (2024), this study is particularly relevant to our work because it demonstrates how language models can analyze and generate arguments, bridging computational methods with theory-driven argumentation frameworks.

In sum, while Large Language Models offer promising tools for argumentation mining, including tasks related to pathos and persuasive discourse, their limitations must be clearly understood. Their outputs, though often convincing, are the result of statistical pattern-matching rather than reasoning or communicative intent. Their utility lies in supporting, not replacing, human interpretation, especially in tasks that benefit from human-annotated prompts or domain-specific knowledge. As this section has shown, LLMs can help identify, generate, and classify argumentative structures, but they must be employed within theory-informed frameworks and accompanied by critical awareness of their constraints. The following chapters develop this perspective, presenting a model for using LLMs in pathos mining that integrates rhetorical theory with computational annotation.

7.4 Large Language Models in Studies on Emotions in Language

Moving on from argument mining, this section turns to a closely related but less explored area: the application of LLMs in studying emotions in language. Although emotion recognition has long been part of affective computing, the incorporation of rhetorical categories such as pathos into automated analysis is still limited. This section aims to identify the specific gap that will later be addressed through the concept of pathos mining. Still, the pathos mining concept is based on studies reviewed here, as they demonstrate how LLMs can support or even participate in the annotation of emotionally charged language.

One of the most relevant studies for the research presented here is that by Moon et al. (2024). In their work, the authors explore the use of logos, ethos, and pathos as rhetorical strategies by fine-tuning GPT-3.5 to automatically identify these strategies within a Korean Q&A database on health-related issues, primarily those related to the COVID-19 pandemic. Their research focuses on determining which rhetorical strategy is perceived by users as the most effective. This is operationalized by identifying the answers that are the most popular among users. Their findings reveal that in their corpus, users show a clear preference for logos-based responses. The paper by Moon et al. reports a high level of inter-annotator agreement (Cohen's Kappa = 0.8), which is noteworthy given that the annotators were the authors themselves. This high agreement is likely facilitated by the use of a straightforward and transparent annotation scheme tailored to the categories present in their dataset. However, the annotation scheme appears less rooted in theoretical frameworks and more derived from the specific features of the corpus.

The primary contribution of Moon et al.'s study to the research presented here lies in demonstrating the benefits of corpus-specific annotation schemes. While such schemes may offer limited theoretical generalization ability, they often lead to higher inter-annotator agreement. Moreover, their findings suggest that a stable, well-annotated corpus can serve as a foundation for scaling rhetorical strategy identification through prompt engineering and fine-tuning. As the authors emphasize, this study not only illustrates the feasibility of applying large language models (LLMs) in humanities and social science research but also expands the potential applications of AI-driven text analysis.

In Moon et al., users show a clear preference for responses based on logos. This outcome contrasts with our findings reported in Part II of this book. In our experimental research, users rated messages containing positive pathos appeals as the most persuasive, as measured by the Perceived Argument Strength (PAS) score. This contrast points to the importance of contextual factors in the effectiveness of rhetorical strategies. Specifically, the subject matter (university exams versus health-related concerns) appears to influence users' preferences. In addition, the operationalization of logos, ethos, and pathos can also play a significant role in shaping the results. Cultural differences must also be considered, as Moon et al.'s study examines Korean internet users, whereas our study had participants speaking Polish. These factors point to the need for future research to adopt an interdisciplinary approach, combining corpus analysis with psychological methods, while also addressing cross-topic and cross-cultural variations.

Another interesting study explores the integration of GPT-4 into the annotating team for the complex task of Entity-Level Sentiment Annotation (ELSA, Rønningstad et al. 2024). The authors employ a detailed annotation scheme that annotates the overall sentiment expressed in the text toward each volitional entity. This high-quality scheme, already validated and supported by a dataset previously annotated for similar purposes, forms the basis of their work. Crucially, this task exemplifies a challenge in human-level annotation due to its complex nature.

To clarify, consider a simplified definition of negative sentiment, such as identifying sentences containing the negation of the verb "to like". Such a task could be taught to undergraduate students in minutes and easily implemented in a machine learning model, but utterly useless. Researchers in argumentation and related fields aim to uncover deeper, complex textual relationships. Using corpus-based methods, they seek to analyze large-scale textual patterns, such as emotional appeals in pathos mining, in ways that reflect human perception. Similarly, Rønningstad et al. (2024) investigate the intricate task of estimating the sentiment level of an entity, focusing on how human perception of global sentiment toward an entity is perceived and how this process can be automated.

Their findings are relevant to our discussion. They observed that while GPT-generated labels often deviate from the majority of human annotators, there is usually at least one human annotator whose labels deviate more than those of the model. This suggests that while LLMs should not be uncritically trusted, human annotators are also not infallible. The authors conclude that LLMs can be valuable contributors to annotation teams, but recommend that their outputs be compared with human annotations or curated by expert reviewers. We will follow this path in studies that include LLMs and will be presented in later chapters of this book.

Similarly, in psychological studies based on natural language texts, GPT models are increasingly recognized as valuable tools for annotation (Rathje et al. 2024). This recognition is based on the growing assumption in psychology that text analysis can provide insights into cognitive and emotional processes without the need for direct involvement of human participants (Jackson et al. 2022). In Rathje et al. (2024), the authors extend this idea by applying modern technology to explore automated analysis of texts that have not been previously available online. This is significant as it ensures that the testing data have not been seen by the model before. A similar approach is applied in my studies presented in Chap. 9, where the data are transcribed by my team or sourced from recordings of argumentation elicitation sessions. Given the increasing scarcity of new training data for large language models (Jones 2024), this method of generating novel data from experimental settings represents a promising direction.

Returning to Rathje et al. (2024), the authors demonstrate that GPT models can effectively identify textual expressions of psychological constructs, such as sentiment, discrete emotions, offensiveness, and moral foundations. Furthermore, unlike typical annotators, the models can perform these tasks in 12 languages, including the less commonly spoken ones. The authors also conducted test-retest assessments to evaluate the consistency of the model over time, finding high reliability in both measurements. Although the models are not flawless, their performance is comparable to that of annotators with minimal training and exceeds that of lexicon-based approaches.

Together, these studies point to the emerging role of LLM as both tools and topics of inquiry in the analysis of emotional language. They offer methodological insights, such as the value of corpus-specific annotation schemes, the potential of LLMs to support human annotators, and the reliability of model-based emotion detection across languages, that are directly relevant to the development of automated approaches to pathos. At the same time, they show the importance of grounding computational strategies in theoretically meaningful distinctions and context-sensitive data.

In the following sections, I expand on this foundation by proposing a framework for pathos mining. This framework is based on the Model of Interactional Pathos in Argumentation, introduced in Part I, and draws on rhetorical theory, cognitive science, and corpus-based methods to systematically identify and interpret emotional appeals in argumentative discourse.

References

Bonaldi H, Damo G, Ocampo NB, Cabrio E, Villata S, Guerini M (2024) Is safer better? the impact of guardrails on the argumentative strength of LLMs in hate speech countering. In: Proceedings of the 2024 Conference on Empirical Methods in Natural Language Processing (EMNLP), Association for Computational Linguistics, Online and Singapore, pp 3446–3463, https://aclanthology.org/2024.emnlp-main.292

Carrasco-Farré C (2024) Large language models are as persuasive as humans, but how? about the cognitive effort and moral-emotional language of LLM arguments. arXiv preprint arXiv:2404.09329, https://doi.org/10.48550/arXiv.2404.09329, https://arxiv.org/pdf/2404.09329, version 2, submitted on April 21, 2024, 2404.09329

Chang Y, Wang X, Wang J, Wu Y, Yang L, Zhu K, Chen H, Yi X, Wang C, Wang Y, et al. (2024) A survey on evaluation of large language models. ACM Transactions on Intelligent Systems and Technology 15(3):Article 39, 1–45, https://doi.org/10.1145/3641289, https://dl.acm.org/doi/10.1145/3641289

Choudhury S, Choudhury A, Kan MY (2021) Argument mining with transformer models. In: Proceedings of the ACL 2021 Workshop on Argumentation, pp 1–10

Falk N, Waldis A, Gurevych I (2024) Overview of perspectivearg2024: The first shared task on perspective argument retrieval. In: Proceedings of the 11th Workshop on Argument Mining (ArgMining 2024), Association for Computational Linguistics, pp 130–149

Haidt J, Joseph C (2004) Intuitive ethics: How innately prepared intuitions generate culturally variable virtues. Daedalus 133(4):55–66, https://www.jstor.org/stable/20027945, open Access, On Human Nature (Fall, 2004)

Jackson JC, Watts J, List JM, Puryear C, Drabble R, Lindquist KA (2022) From text to thought: How analyzing language can advance psychological science. Perspectives on Psychological Science 17(3):805–826, https://doi.org/10.1177/174569162110048

Jones N (2024) The ai revolution is running out of data. what can researchers do? Nature 636:290–292, https://doi.org/10.1038/d41586-024-03990-2, news Feature

Kocoń J, Cichecki I, Kaszyca O, Kochanek M, Szydło D, Baran J, Bielaniewicz J, Gruza M, Janz A, Kanclerz K, Kocoń A, Koptyra B, Mieleszczenko-Kowszewicz W, Miłkowski P, Oleksy M, Piasecki M, Łukasz Radliński, Wojtasik K, Woźniak S, Kazienko P (2023) Chatgpt: Jack of all trades, master of none. Information Fusion 99:101861, https://doi.org/10.1016/j.inffus.2023.101861

Landowska A, Budzynska K, Zhang H (2024) Quantitative and qualitative analysis of moral foundations in argumentation. Argumentation 38:405–434, https://link.springer.com/article/10.1007/s10503-024-09636-x

Lawrence J, Reed C (2015) Combining argument mining techniques. In: Proceedings of the 2nd Workshop on Argumentation Mining, Association for Computational Linguistics, Association for Computational Linguistics, Denver, Colorado, pp 127–136, https://aclanthology.org/W15-0516.pdf

Lawrence J, Reed C (2019) Argument mining: A survey. Computational Linguistics 45(4):765–818

Levy R, Bilu Y, Bogin D, Eden M, Sznajder B, Slonim N (2014) Context-aware argumentative mining in online discussions. In: Proceedings of COLING 2014, Association for Computational Linguistics, pp 883–892

Moon H, Bae BJ, Bae S (2024) Developing a chatgpt-based text extraction model to analyze effective communication elements in pandemic-related social q&a responses. In: 2024 International Conference on Artificial Intelligence in Information and Communication (ICAIIC), IEEE, IEEE, Seoul, Republic of Korea, pp 1–6, https://doi.org/10.1109/ICAIIC60209.2024.10463227

Rathje S, Mirea D, Sucholutsky I, Marjieh R, Robertson C, Bavel JV (2024) Gpt is an effective tool for multilingual psychological text analysis. Proceedings of the National Academy of Sciences of the United States of America (PNAS) 121(34):e2308950121, https://doi.org/10.1073/pnas.2308950121

Rønningstad E, Velldal E, Øvrelid L (2024) A gpt among annotators: Llm-based entity-level sentiment annotation. In: Proceedings of The 18th Linguistic Annotation Workshop (LAW-XVIII), Association for Computational Linguistics, University of Oslo, Language Technology Group, pp 133–139

van Rooij I, Guest O, Adolfi F, de Haan R, Kolokolova A, Rich P (2024) Reclaiming ai as a theoretical tool for cognitive science. Computational Brain & Behavior Open access

Ruiz-Dolz R, Lawrence J, Reed C (2024) Overview of dialam-2024: Argument mining in natural language dialogues. In: Proceedings of the 11th Workshop on Argument Mining (ArgMining 2024), Association for Computational Linguistics, pp 83–92

Şahinuç F, Kuznetsov I, Hou Y, Gurevych I (2024) Systematic task exploration with LLMs: A study in citation text generation. In: Proceedings of the 62nd Annual Meeting of the Association for Computational Linguistics (Volume 1: Long Papers), Association for Computational Linguistics, Bangkok, Thailand, pp 4832–4855, https://doi.org/10.18653/v1/2024.acl-long.265

Saha S, Srihari R (2024) Turiya at dialam-2024: Inference Anchoring Theory Based LLM Parsers. In: Proceedings of the 11th Workshop on Argument Mining (ArgMining 2024), Association for Computational Linguistics, Bangkok, Thailand, pp 124–129, https://doi.org/10.18653/v1/2024.argmining-1.13

Stab C, Gurevych I (2017) Parsing argumentation structures in persuasive essays. Computational Linguistics 43(1):619–659

Woźniak S, Koptyra B, Janz A, Kazienko P, Kocoń J (2024) Personalized large language models. In: Proceedings of the SENTIRE 2024 (ICDM Workshops), https://doi.org/10.48550/arXiv.2402.09269, accepted to SENTIRE 2024 (ICDM Workshops), 2402.09269

Chapter 8
Lexicons and Models for Pathos Mining

Abstract This chapter presents computational methods for analyzing emotional appeals in natural language argumentation using psychological affective lexicons and sentiment analysis. Building on earlier parts of the book, it operationalizes the Model of Interactional Pathos in Argumentation (MIPA) by identifying emotion-eliciting words in argumentative structures and modeling audience reactions. Using Polish and English pre-election debates, the chapter demonstrates how lexicon-based methods capture emotional appeals in premise-conclusion pairs and how sentiment analysis can detect audience reactions in real-time social media commentary. It introduces a threefold methodological framework: lexicon-based pathos mining, audience response sentiment analysis, and manual annotation of emotional appeals in online interactions. The results indicate strong correlations between emotional appeals and audience sentiment, including unexpected cross-emotional patterns (e.g., anger appeals that generate sadness). A study of Twitter discussions shows that while negative emotional appeals dominate, positive pathos increases engagement and can soften antagonistic responses. The chapter argues that lexicons and sentiment models, though limited in context sensitivity, offer scalable interdisciplinary tools for the empirical study of rhetorical emotion in discourse. It lays the foundation for subsequent chapters on generative models for pathos mining.

Keywords Pathos mining · Affective lexicons · Sentiment analysis · Audience response modeling · Argument mining · Emotional appeals

8.1 Using Lexicon-Based Methods for Pathos Mining

In this chapter, I will discuss how pathos can be identified in natural language argumentation by detecting emotion-eliciting words in premise-conclusion pairs, a technique which can be called pathos mining with lexicons (per analogy to argumentation mining with lexicons). As a case study, I will apply the psychological affective lexicons introduced in Part II of this book to the corpus of pre-election debates presented in Part I. This analysis builds upon work originally conducted for the paper by Konat et al. (2024), however, here it is developed into a full

B. Konat, *Emotional Appeals in Argumentation*, Argumentation Library 47,
https://doi.org/10.1007/978-3-032-06332-8_8

operationalization of the Model of Interactional Pathos in Argumentation (MIPA), as presented in Part I of this book.

Affective lexicons are tools developed by psychologists to assess the emotional impact of lexical stimuli (i.e., words) on an average human respondent. These assessments are based on established lists of words (Mohammad and Turney 2013; Wierzba et al. 2015, 2021). Lexicons with emotional norms for various words in a given language were originally created by psychologists to help control emotional valence (positive or negative), arousal, and basic emotions during the creation of experimental stimuli, as described in Part II of this book. However, these lexicons can also be applied in computational linguistics to enrich manual or automatic annotation of argument structures. This can be stored in formats such as AIFdb (Lawrence et al. 2012) and obtained through automated mining methods. To incorporate emotional appeals (pathos) into argument structure, text from each element (e.g., premises, conclusions) is tokenized, and relevant words are matched with entries in the lexicon. This process highlights emotion-eliciting words and their associated valence, arousal, and specific emotions (such as fear, joy, anger, disgust, or sadness), depending on the lexicon used.

In this section, I will integrate these two elements: a list of emotion-eliciting words from affective lexicons and the transcripts of pre-election debates, to examine how speakers use such words in their premises and conclusions. The goal is to observe how these words are used to evoke emotions in the audience, enhancing the persuasiveness of the arguments.

In the final chapter of Part I, I demonstrated how manual identification (annotation) of argument scheme usage can reveal pathos strategies in natural language. Here, I will employ a computational approach using a simple method described in the paper by Konat et al. (2024). This algorithm compares the words in the premises or conclusions of the arguments with those in the affective lexicon. When a match is found, the associated emotional values are recorded in a table. This straightforward yet effective method identifies emotion-eliciting words used by speakers. It serves as a complementary technique to the manual annotation method discussed in Part I and is the first of several computational methods introduced throughout Part III.

Although much of the existing work in argumentation theory addresses emotional appeals through terms such as "fear" or "joy", these concepts are often treated as intuitively understood by both theorists and practitioners. By incorporating psychological knowledge, specifically lexicons that systematically link words to emotions, I propose a framework that complements traditional approaches. This interdisciplinary perspective enriches argumentation theory by aligning with Aristotle's concept of pathos, providing empirical tools for understanding how emotions are elicited in the audience. Furthermore, the application of computational linguistics, such as matching emotional words from psychological lexicons to arguments in databases like AIFdb, offers a novel and practical method for analyzing emotional appeals in natural language argumentation.

In a study conducted by my team, presented in Konat et al. (2024) we applied this method to four pre-election debates. We focused on the argumentative sections of the debates, selecting portions where logos-based argumentation had been marked using Inference Anchoring Theory (IAT). For Polish, we used the Emotion Meanings lexicon and the Nencki Affective Word List (NAWL). For English, we selected the NRC Word-Emotion Association Lexicon (EmoLex). The intensity of emotion-eliciting language was computed for each argument unit using the average emotional intensity of the lexicon's categorized words. Our findings show that nearly 95% of the arguments contain emotion-eliciting words, with an average of 2.75 such words per unit of argument (min = 0, max = 23).

The presence of emotion-eliciting words in premise and conclusion pairs can be observed in examples taken from the pre-election debates.

(1) Andrzej Duda (May 6, 2022; TVP):
Premise: Adoption by same-sex couples is forbidden by the Polish **constitution**.
Conclusion: Adoption by same-sex couples is **absolutely** out of the question.[1]

(2) Wladyslaw Kosiniak-Kamysz (June 17, 2022; TVP):
Premise: We should appreciate **hard-working** people by raising the tax-free amount.
Conclusion: We should not **punish** entrepreneurs.[2]

In examples (1) and (2), the words in bold, identified in the lexicons illustrate how politicians use both positively valenced words (e.g., "constitution", "hard-working") and negatively valenced words (e.g., "absolutely", "punish") in constructing arguments that appeal to different emotions. In the persuasive context of pre-election debates, the speakers are constructing argumentation to either support the practical reasoning call to action ("vote for me") or to provide reasons for their decisions and beliefs. This second case is illustrated in example (1), where the current president, Andrzej Duda, who is associated with conservative views and maintains close ties to the Catholic Church, argues against allowing same-sex couples to adopt children. As a reason for that, he claims that this act is forbidden by the Polish constitution. The word "constitution" usually elicits positive valence (0.63 on average) and mostly happiness (2.96), with some elements of fear, however (2.08), according to the lexicon (Wierzba et al. 2015).

In example (2) the speaker uses the negatively valenced word "punish" with negation, to show that increasing taxes for corporation is, in fact, punishing entrepreneurs. According to the study (Wierzba et al. 2015), the word "punish" has a strongly negative valence (−1.33) and tends to elicit mainly anger (4.07), sadness (4.07) and fear (4.3). The speaker is skillfully including this word in his conclusion, to appeal to the anger of this audience—we do not want to punish entrepreneurs, why should we? This allows him to use pathos for persuasive gain, by hiding other aspects of tax increase (better budget for social care, for example) and to highlight negative ones, and frame them as "punishment". Here we can see how pathos is creating additional layer to logos. Although propositions can be structured into

[1] [Pol.] *Adopcja przez pary jednopłciowe jest zakazana przez polską konstytucję. Adopcja przez pary jednopłciowe jest absolutnie wykluczona.*

[2] [Pol.] *Trzeba doceniać ciężko pracujących ludzi, podnosząc kwotę wolną od podatku. Nie można karać przedsiębiorców.*

premises and conclusions (as they have been collected from AIFdb (Lawrence et al. 2012)), the pathos layer is intertwined on the linguistic surface, emphasizing how the two persuasive means are related in natural language exchanges.

(3) a. Premise: Over the **last** few months, [Andrzej Duda, the president] together with Mateusz Morawiecki have done everything to lower **wages** and **lay off people**.[3]
b. Conclusion: (therefore) it amuses me when Mr. Andrzej says that higher wages will be a **condition** for adopting the Euro.[4]

In example (3), the speaker smoothly includes a combination of positive and negative emotive words in his argument, to increase the persuasive power of this argument. Especially striking here is the combination of positively valenced "wages" and "people" with the "lay offs" and the "lowering", which obviously aims at creating contrast and emotional seesaw. All this aims at the character attack expressed in the conclusion, in the ironic form emphasized by the informal and very inappropriate form of addressing the president with his first name, as "Mr Andrzej". We might assume that the speaker does it on purpose, to attack the character of the current president and undermine his ethos.

(4) a. Premise: The very **fact** that we will be treated like farm **animals**, that we can be **forcibly** vaccinated or not, that our **child** can be taken away and vaccinated. This fact alone is **terrible**, **monstrous**.[5]
b. Conclusion: (therefore) it does not matter whether this vaccine will be effective, whether it will exist or not.[6]

Example (4) presents a cornucopia of emotive words used in such an over-dramatic manner that the argument verges on the parody, thus maybe missing its goal of eliciting fear. The cumulation of emotive words in the premise (including "animals", "forcibly", "terrible", 'monstrous") contrasting with the vision of taking away the child is contrasting with a very calm and down-to-earth claim in the conclusion. The speaker is arguing that it does not matter whether the Covid vaccine would be effective or not, as what really matters is the obligatory aspect of vaccination. Instead of maybe basing it on legal or ethical considerations, the speaker is instead appealing to emotions with a rich comparison of humans to farm animals, forcibly vaccinated. The use of lexicons allows us to automatically identify such words and thus can support further automatic or manual analysis of emotional appeals.

Of course, this method has significant limitations, as it is context-insensitive. For this reason, it should be complemented by other methods, either experimental or computational. Let us examine an example (5) in which, through a simple bon

---

[3] [Pol.] *Przez ostatnie miesiące, [Andrzej Duda, prezydent] razem z Mateuszem Morawieckim zrobili wszystko, żeby wynagrodzenia obniżyć i zwalniać ludzi.*

[4] [Pol.] *(zatem) śmieszy mnie wypowiedź Pana Andrzeja, który mówi, że wypłaty wyższe będą warunkiem przyjęcia Euro.*

[5] [Pol.] *Sam fakt, że będziemy traktowani jak zwierzęta gospodarcze, że można nas zaszczepi´ siłą albo nie, że można nam dziecko odebrać, zaszczepić. Już sam ten fakt jest straszny, potworny.*

[6] [Pol.] *(zatem) nieważne czy ta szczepionka będzie skuteczna, czy ona będzie, czy nie.*

mot, a speaker with a liberal orientation argues that lower taxes will result in a less expensive state:

(5) a. Premise: There will be lower **taxes**.[7]
b. Conclusion: (therefore) the **state** will be less **expensive**.[8]

Unfortunately, the word "taxes" is marked in the lexicon (Wierzba et al. 2021) as eliciting a strong anger. Of course in this context, the speaker is talking about lowering taxes; however, the method is not sensitive to it. Therefore, we will move to other methods of pathos analysis, which are more sensitive to context; however, they lack the precision and solid psychological foundations of affective lexicons.

The lexical method, although simple, allows for a large-scale automated analysis of emotional appeals in argumentation. However, it has limitations, such as omitting words not listed in lexicons but likely to elicit emotions. Another challenge is the Pollyanna effect (Boucher and Osgood 1969), where positive language is disproportionately represented due to the human tendency to use positive words rich in value. Despite these challenges, the method offers a useful tool for large-scale analysis of emotional appeals in natural language argumentation.

8.2 Modeling Audience Reactions with Machine Learning

Let us now return to the study of argument schemes as a form of emotional appeal, presented in Part I of this book (Chap. 3). Since both types of emotional appeal, through argument schemes and through emotional language, have now been discussed, we can turn to the audience's side of the interaction.

Figure 8.1 presents an operationalization of the Model of Interactional Pathos in Argumentation (MIPA), this time focused on how audiences respond to emotional appeals in televised debate settings. The figure illustrates the two cognitive and linguistic layers involved: on the speaker's side, emotional appeals are performed using pathotic argument schemes; on the audience's side, emotional reactions are observable in the form of emotion-expressing language. The unidirectional arrow highlights the rhetorical flow from speaker to audience, and the inclusion of social media icons points to the relevance of such audience expressions in online settings.

One of the methods for capturing audience reactions expressed in text is through the use of sentiment analysis models from computational linguistics. A key distinction to reiterate here is between eliciting and expressing emotions. While psychological lexicons, mentioned earlier, identify words that usually elicit emotions in humans when heard, sentiment analysis models identify emotions expressed in text. This distinction, of different perspectives of the speaker and the

[7] [Pol.] *Będą mniejsze podatki.*

[8] [Pol.] *(Zatem) będzie tańsze państwo.*

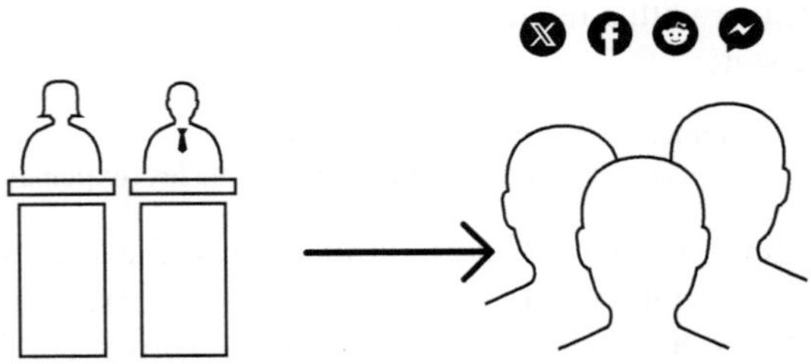

Fig. 8.1 Model of Interactional Pathos in Argumentation (MIPA): emotion-expressing language

audience, is gaining recognition in both the sentiment analysis community (Buechel and Hahn 2017; Alsaedi et al. 2022) and the argumentation research community (Lukin et al. 2017). As presented in Part I of this book this distinction is crucial for the Model of Interactional Pathos in Argumentation. For example, words like "children" or "hurricane" do not express emotions but may evoke them. In contrast, an expressed emotion can be seen in a consumer review stating, "I do not like this product," and sentiment analysis models, typically trained on corpora of such statements, are ideal tools for capturing audience reactions to emotional appeals in text.

Currently, there is a wide range of sentiment analysis models for various languages (Nakov et al. 2019; Nguyen et al. 2020; Kocoń et al. 2021; Barbieri et al. 2022; Hartmann 2022), and they are widely used to analyze natural language corpora (Bednarek 2008; Antypas et al. 2022). Selected researchers have already explored the usefulness of sentiment analysis for argumentation studies (Basile et al. 2016; Cabrio and Villata 2018; Stede 2020). This topic is particularly important in the context of propaganda and hate speech detection (Yoosuf and Yang 2019; Da San Martino et al. 2019).

In our study, presented in Konat et al. (2024), we used the pre-election debates corpus mentioned in Chap. 3, and in this corpus we have identified emotional appeals with manual identification of argument scheme instances (as described in Chap. 3) and by automated identification of emotion-eliciting words in argument structure (as described in Chap. 8).

But in this study, we wanted to look at interactional model of pathos (MIPA) and fully incorporate both sides—an emotional appeal to emotion on the side of the speaker and emotional response on the side of the audience. One way to capture audience reactions is to perform psycholinguistic experiments, as presented in Chap. 6. Another way to capture audience reactions is to observe emotions expressed by the audience in direct, time-aligned responses to emotional appeals. In the study presented in Konat et al. (2024), we treated the response in real-time social media to debates as a valuable indicator of audience emotions. To analyze emotions expressed by the audience in responses to emotional appeals, we used corpora comprising Reddit comments related to the US primary debates, available on the AIFdb platform (US2016R1reddit and US2016G1reddit corpora) (Visser et al. 2020). To measure emotional responses in Polish data, we collected comments written on X (formerly Twitter) during live presidential debates.

By aligning the timing of the comments with the debates, we obtained a statistically significant result showing a relationship between the saturation of emotion-eliciting and emotion-expressing language. Specifically, the more politicians appealed to emotions, either through argument schemes or emotion-eliciting words, the more emotionally strong will be the audience's responses.

Analysis of expressed sentiment revealed that, in general, in social media reactions to political debates, speakers expressed mainly negative sentiment (58%) compared to lower proportions of neutral (28%) and positive sentiment (14%). When we examined categorical emotions, the dominant was surprise (58%), followed by anger (18%) and joy (12%). We interpret the dominance of surprise as a specific feature of the corpus which comprises direct reactions to politicians' speeches. Speakers are likely to express their surprise (often with a negative connotation) at what they have heard. At the same time, we should approach results related to the emotion of surprise with caution. As discussed in Chap. 4, this emotion is often unclear to respondents and annotators. Consequently, trained models may also produce unreliable responses, as the accuracy of a model depends on the quality of the data used to train it.

How do these overall results translate into reactions to specific emotional appeals? First, we analyzed fluctuations in expressed emotions after a politician used an argument scheme categorized by us as appealing to emotions, selected from the classification of Walton et al. (2008). As described in Part I, those are: Argument from Positive Consequences, Argument from Negative Consequences, Argument from Fear Appeal, Argument from Danger Appeal, Argument from Threat, Slippery Slope Argument, Argument from Waste, and Argument from Need for Help. Using our annotation of argument schemes presented in Chap. 3, we aligned the use of each scheme with time-stamped social media data and measured correlation. This analysis yielded two statistically significant results. The first reveals that in one of the English-language debates, appeals to negative emotions using the Waltonian (2008) scheme of Negative Consequences corresponded with expressions of negative sentiment. For one of the Polish debates, we found an even stronger correlation: after an emotional appeal employing the Negative Consequences scheme, social media audiences expressed statistically significantly more anger.

This suggests that while the argument scheme remains consistent across languages, the emotional expression and its intensity vary. Moreover, this finding points to the differences in the quality of the models for Polish and English, which complicates direct comparisons between the two. This leads us to the usefulness of multilingual models, as we shall discuss in forthcoming chapters.

In the study discussed earlier in this chapter, we looked at a second type of emotional appeal in language: the use of emotion-eliciting words, as cataloged in the affective lexicons outlined in Chap. 4. When we matched the appearance of emotion-eliciting words in argumentative structures from pre-election debates with social media reactions, we found nine statistically significant relationships. A particularly strong correlation was observed between anger-eliciting words in argument premises and sadness expressed by the audience ($r = 0.77$). Three weaker correlations were noted: anger-eliciting words in argument premises with expressed joy ($r = 0.27$); fear-eliciting words in argument premises with expressed joy ($r = -0.29$); and sadness-eliciting words in argument conclusions with expressed fear ($r = 0.26$). This shows a complex network of relationships: when speakers use words aimed at eliciting anger, intending to provoke this emotion and potentially motivate action (since anger is an energetic emotion linked to action, as defined in Part II of this book), the audience may respond in an unexpected way. This shows the importance of context in emotion-eliciting language. Words that elicit specific emotions in a context-free setting, such as during the construction of emotional lexicons (as described in Part II of this book), may evoke entirely different responses when used in the context of political speech, where additional factors influence their impact.

The key general finding of this study is that emotional appeals and emotional reactions expressed in language form observable pairs that can be analyzed using corpus methods. This provides fascinating insights into the dynamics of speakers' appeals and audience reactions, matched in sequential orders. Unlike the experimental methods presented in Part II, this method does not create an artificial environment; instead, it allows for the observation of arguments and reactions in natural contexts. In experimental settings, as described in Chap. 6, participants who completed questionnaires tended to yield to social pressure, responding in ways that they believed researchers wanted to see or in ways they wanted to perceive themselves. For example, participants often indicated that positive emotional appeals were more persuasive. This finding is counterintuitive, as real-world observations in scams, dark persuasion, and political campaigns show that negative emotional appeals are perceived as powerful and are becoming increasingly prevalent. This illustrates how psycholinguistic and corpus methods are complementary: the former provides controlled insights into contributing factors, while the latter, despite results being obscured by numerous variables, offers a more realistic understanding of how people react to emotional appeals and which emotions they express in response.

8.3 Model of Interactional Pathos in Argumentation in Social Media Discourse

In the previous section, we examined the unidirectional communicative setting of pre-election debates, where politicians act as rhetors deploying pathos, and audiences respond with emotion-expressing language. However, these dynamics are different in on-line environments, where each participant is potentially both rhetor and audience. In social media discourse, users respond to appeals while

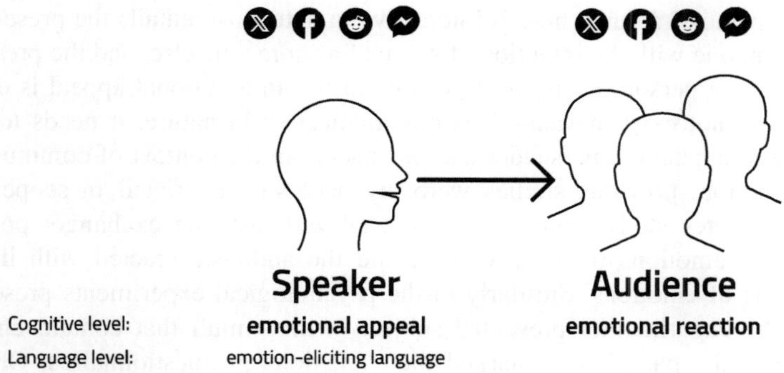

Fig. 8.2 Model of Interactional Pathos in Argumentation (MIPA): speaker–audience role fluidity in social media argumentation

simultaneously producing their own, often in everyday, informal language rather than institutionalized rhetorical speech.

Figure 8.2 proposes another operationalization of the Model of Interactional Pathos in Argumentation (MIPA), which conceptualizes this dual role and illustrates how emotional appeal-response sequences can be captured using corpus-based methods. The figure consists of two stylized human profiles of speaker and audience, and above both figures, icons representing popular social media platforms signal that this model applies specifically to digital communication environments. Importantly, the audience is represented as a group of overlapping silhouettes, to represent the multiplicity and anonymity of recipients in social media contexts. This operationalization of MIPA serves as a foundation for the study of pathos in online interaction presented in this section.

Online discussions, even though they are seemingly similar to everyday spoken dialogues, seem to be governing with additional special rules. In my opinion, an interesting insight is provided by Douglas Walton, when long before the advent of social networks he writes about a concept of "group quarrel" which "is built around the rhetoric of group belonging and emotional opposition to another group with whom the first group is, or is alleged to be, in conflict" (Walton 1992, p. 273). This idea aligns well with the concept of in-group and out-group in social psychology—also long before social media, where individuals identify strongly with their own group (in-group) while often exhibiting antagonism or bias against those perceived as outsiders (out-group) (see, e.g., Tajfel and Turner 1986).

In previous sections of this chapter I described automated methods and results which I personally use and find effective in analyzing emotions in argumentation: matching words from psychological lexicons into natural language argumentation structures to capture appeals to emotions, pathos, and using sentiment analysis models to capture audience reactions to those appeals. In this section I would like to say more about different combinations of methods that can provide us with more insight into the interactive nature of pathos. As presented in Part I of this book, the model of pathos I assume here is interactive in nature—it entails the presence of a rhetor, someone with the intention of persuading someone else, and the presence of an audience, a person or group of persons to whom emotional appeal is directed. Pathos does not exist in vain—it is communicative in nature, it needs to have a speaker and a hearer, a persuader and persuadee. In the context of communication and interaction, previous studies were, by necessity, restricted in scope. In the political debates study from Part I we analyzed just one exchange: politicians appealed to emotions of the audience, and the audience reacted with linguistic expressions of emotions. Similarly in the psychological experiments presented in Part II: the experimenter presented subjects with stimuli that contain emotional appeal, and the participants marked their reactions in questionnaires. Observing multiple dialogical turns in natural language argumentative exchanges—where one emotional appeal prompts a reaction, potentially followed by another emotional appeal—remains a goal for the future. However, the study presented in this section represents a meaningful step toward understanding the interactional pathos. This work attempts to bridge the gap between rhetorical theory and computational analysis. By approaching pathos as an interactive, dialogical process rather than a static rhetorical device, this study aims to reveal how emotional appeals evolve through speaker-audience exchanges, offering a model that can shed new light on the dynamics of public discussions and digital communication. As a material here, we selected social media interactions. In social media each user can take the role as both rhetor and audience in a dynamic, interactive exchange. In any given interaction, users may alternate between appealing to the emotions of others and responding to others' emotional expressions. As a result, understanding the pathos on social networks requires a model that accounts for the constant exchange and mutual influence of emotions across posts and responses.

The study on pathotic interactions in public discussions about social issues on Twitter was published in Gajewska and Konat (2024). The material for this original study was drawn from annotated corpora addressing polarization issues in online group discussions, as published in the previous study (Gajewska et al. 2024). The data set spans five collections that cover topics like COVID-19 vaccines, climate change, and the 2016 US presidential election, sourced from Reddit and Twitter. Discussions on COVID-19 and climate change were collected from specific Reddit groups, such as r/conspiracy and r/climatechange, while related Twitter data was gathered through keyword searches in initial tweets and their replies. For the study in Gajewska and Konat (2024), we selected only the pathos annotation layer, which was performed under my supervision by 8 students using the pathos annotation scheme prepared specifically. Going back to the previous manual annotation study presented in this book (in Part I), we now have two different approaches and methods of capturing pathos with manual annotation in corpora presented here: manual annotation of argument schemes presented in Part I of this book and manual annotation of sentences as containing pathos, which is used here. For pathos annotation prepared for Gajewska et al. (2024), we decided not to apply argument schemes this time, but for annotators to go sentence by sentence and decide in the first step, whether the speaker has intention to elicit emotions in hearers for rhetorical gain, and if yes, whether this emotion is intended as positive or negative. This resulted in marking about 30% of the whole dataset as containing pathos.

The resulting corpus is rich with examples of appeals of natural language to emotions. Users are trying to encourage each other to get vaccinated:

(6) jblank84: The good news is that you can, friend!

Or to plant trees:

(7) dolmakilroy: My parents literally planted thousands of trees afterwards, which we must continue to do.

As we can see, users are engaged in persuading others with a practical reasoning scheme, introducing positive emotional appeals aimed at eliciting positive emotions in the audience. They do this either by emphasizing possibility and fostering a sense of warmth and phatic connection (e.g., "you can, friend") or by first presenting a heartwarming image of parents planting thousands of trees and then following it with a call to action ("which we must continue to do"). The strength of the emotional appeal becomes clearer when we imagine the call to action without it; a message like "We must plant thousands of trees" lacks the convincing effect of the same call preceded by the emotional appeal. Emotional appeal improves the strength of the call to action, even if it does not provide a formal argument or a valid reason for action. This situation resembles that described in Younis's (2023) study on the perceived strength of rephrasing: while rephrasing lacks the persuasiveness of solid arguments, it is more convincing than providing no support at all. Similarly, emotional appeal serves here as a persuasive strategy not adjacent to logos but often in its place, standing in for argumentative support.

This corpus of online exchanges contains a higher proportion of negative appeals—about 80% of all identified pathos instances—compared to positive ones. This bias aligns with the intention to select polarizing topics, with discussions around vaccines in particular intensifying the tendency to use negative valence in calls to action:

(8) NilacTheGrim: It's just supposed to make the people that read it believe some nonsense that serves the interests of the people creating the propaganda.

In the case of climate change users also appeal to negative emotions:

(9) FogartyClimate: Climate change has to be urgently addressed, and failure to make deep cuts to all fossil fuel emissions, especially coal, will condemn many of the world's poorest and most vulnerable to a more dangerous and hotter future.

Here, the user is advocating for cuts in fossil fuel emissions, a difficult and costly decision for many individuals and regions that, as an action, requires effort and commitment. Although scientific evidence supporting this stance is available, the user has opted—perhaps rightly, perhaps not—to appeal to the emotions of the audience instead of citing data. Using emotional terms such as "poorest" and "vulnerable", they may elicit compassion, pity and warmth, in contrast to the stark term "dangerous". All these examples from dataset published in Gajewska et al.

(2024) illustrate the diversity in emotional appeals found on social media that can be uncovered during manual annotation.

While these results are insightful, I need to address the challenges my team encountered in annotating pathos appeals within this dataset. Annotators often confused emotional appeals with expressions of explicit emotion (e.g., "It's a big win on so many fronts."), and at times misclassified certain statements—such as "I physically shuddered when she said that "—as positive pathos. Here, the issue is likely a misunderstanding of the verb "shuddered," which typically signals an involuntary reaction to discomfort, rather than a positive emotional response. Thus, this annotation was incorrect, as it first fails to capture the negative nature of the expression, and second confuses emotional expression with emotional appeal.

Annotation of pathos appeals does pose difficulties; by following the methodology of Koszowy et al. (2022) and working directly with their team, we engaged in iterative refinement and systematic error analysis, ultimately achieving a significantly improved Inter-Annotator Agreement (IAA) score. We achieved an IAA of approximately 0.34 when determining whether a sentence contained a pathos appeal at all, which increased to an average of 0.71 in the subsequent task of categorizing pathos as positive or negative. This improvement demonstrates the feasibility of annotating emotional appeals with sufficient reliability, given the right annotation scheme, training, and dataset. Contrary to a common objection, the task of annotating emotional appeals is not impossible simply because emotional responses vary individually or are highly context-dependent. Psychological lexicons (presented here in Part II) have shown that population-level responses to emotional stimuli can indeed be measured reliably, and manual annotation described in the current section, especially with trained annotators, can reach acceptable levels of agreement. Our findings support the notion that with a well-defined framework and rigorous training, achieving stable annotation of emotional appeals in natural language is possible, even if there is still room for improvement in future studies.

Having established an annotated dataset (as published in Gajewska et al. 2024), we then analyzed pathos interactions in social media. We have integrated manual annotation with sentiment analysis methods, and presented these findings in another study, (Gajewska and Konat 2024). For this study, we were mainly interested in which emotions—positive or negative—social media users appeal to in their posts, and how these emotions are expressed in responses. To achieve this, we used the existing manual annotation of emotional appeals and, to capture emotional expressions, added an automated layer of sentiment analysis. On this dataset, we conducted a series of analyses focused on appeal-response interactions to understand how pathos works in social media.

First, our descriptive analysis revealed—perhaps unsurprisingly to anyone familiar with social media discussions—that our dataset is dominated by negativity. As noted earlier, human annotation showed that about 80% of all annotated pathos instances appealed to negative emotions. Our additional automated analysis of expressed emotions confirmed this pattern, emphasizing a strong prevalence of negative emotions. Specifically, fear is the most common category (34.6%), followed by anger (15.6%), sadness (14.7%), surprise (13.7%), joy (10.8%), and disgust (0.2%).

Second, we analyzed direct responses to pathos appeals. To recap, pathos appeals were manually annotated, while responses were automatically analyzed for expressed emotions using sentiment analysis model. For this, we match user's post pairs: identifying a "previous" tweet with a pathos category (none, positive, or negative) and a "following" tweet, where we tracked expressed emotions (fear, anger, surprise, sadness, joy, and neutral) to assess their association with pathos-appealing strategies. Our findings show that anger is more likely to appear in responses to tweets with negative or no pathos appeal compared to positive appeals. Joy is most commonly expressed in responses to positive pathos appeals and least after negative appeals. Fear is more frequent in responses to tweets without pathos appeals and—which is a surprising result—with positive appeals. Taking it all together, we may conclude that in general appealing to positive or negative emotions on social media has significant effects. Specifically, while users often express anger and fear, appeals to positive emotions tend to soften these responses, suggesting a promising basis for counter-strategies to reduce hate speech, polarization, and antagonism through positive emotional engagement rather than solely relying on counter-argumentation or moderation.

Our third type of analysis is particularly interesting, as it is performed in a counter-intuitive direction. So far, we looked at the direction of influence corresponding to the interactive pathos model presented in Part I: first, an emotional appeal on the side of the speaker, then an emotional reaction on the side of the hearer. For this particular analysis, we decided to turn the other way around and observed how social media users react to previous expressions of emotion with an appeal to emotion. Even though it is counterintuitive, it is rooted in theoretical insights presented at the beginning of this section, showing how users dynamically switch between the roles of rhetor and audience. In this context, in the current analytical direction, we look at the dialogical move of a user who perceives someone else's expression of negative emotion first and, in the subsequent move, attempts to elicit specific emotions in this person for their own persuasive gain. Our findings reveal that users often employ very empathetic and humane strategies—when faced with someone's expression of fear, they respond with an appeal to positive emotions (positive pathos), probably trying to soothe the previous poster. This suggests that empathy-driven appeals could serve as an important component in reducing antagonism on social media.

Our final analysis reveals how emotional appeals and reactions together influence audience engagement. Here, contrary to findings related to just expressed emotions, we found that when users use positive pathos as a rhetorical strategy, they tend to get higher engagement. In other words, while it might be true that expressing negative emotions can lead to higher engagement, the opposite is true if we focus on how users try to appeal to the emotions of others—to be effective in this area, it is better to use positive pathos strategies.

In summary, the main findings of the study (Gajewska and Konat 2024) suggest that the pathos on social networks is often negative, especially in polarizing discussions. Positive pathos encourages engagement and may reduce hostility, showing potential for moderation strategies. Empathy-driven responses, especially to fear or anger, build on the dialogical nature of social media interactions, where users switch roles and influence each other's emotions.

When interpreting our findings, we relate them to Papacharissi's (2015) concept of "affective publics," which exemplifies how emotional expression fosters collective identity and solidarity within online communities. Our findings suggest that emotional appeals, whether positive or negative, play a significant role in building collective sentiment among users. By integrating Papacharissi's concept with our model, we see that emotional appeals on social media do more than shape individual responses, they also help to reinforce or disrupt community bonds. Specifically, appeals to positive emotions can cultivate empathy and solidarity, while negative appeals can contribute to polarization. Thus, our study not only describes these dynamics but also shows their potential impact on social cohesion in online spaces.

We also draw on the concept of emotional contagion, which describes how individuals subconsciously "catch" and mirror each other's emotions, often through verbal and non-verbal cues (Hatfield et al. 1993). Traditionally studied through facial expressions, behavior, and physiological indicators (Herrando and Constantinides 2021), emotional contagion research has yet to extensively explore large-scale language data from social media. Our study contributes to filling this gap by analyzing how emotions spread and influence social media discourse, showing that emotional dynamics online are both complex and impactful in shaping interactions.

In conclusion, this chapter has demonstrated how computational linguistics methods, particularly lexicon-based pathos mining and sentiment analysis, can further our understanding of emotional appeals in natural language argumentation. By integrating psychological resources such as affective lexicons with argument structure, and by modeling audience reactions through sentiment classification, I have operationalized the interactive dimension of pathos as proposed in the MIPA framework. The empirical findings of my team reveal that emotional appeals and responses are frequent in public discourse, particularly in political and online settings. Lexicon-based methods provide scalable tools for identifying emotion-eliciting language, while sentiment analysis captures how these appeals are received and echoed by audiences. Although both methods face limitations, particularly in terms of contextual sensitivity and granularity, they enable large-scale, corpus-based analyses of emotional dynamics in argumentation. These findings are in line with the need for interdisciplinary approaches that combine psychological theory, rhetorical models, and computational tools. Together, they offer a promising path forward for analyzing the affective dimensions of persuasion, and the following chapter will extend this inquiry by investigating the role of large language models in pathos mining in natural language contexts.

References

Alsaedi A, Brooker P, Grasso F, Thomason S (2022) Improving social emotion prediction with reader comments integration. In: Proceedings of the 14th International Conference on Agents and Artificial Intelligence, SCITEPRESS-Science and Technology Publications

Antypas D, Preece A, Camacho-Collados J (2022) Politics, sentiment and virality: A large-scale multilingual Twitter analysis in Greece, Spain and United Kingdom. arXiv preprint arXiv:220200396

Barbieri F, Anke LE, Camacho-Collados J (2022) Xlm-t: Multilingual language models in Twitter for sentiment analysis and beyond. In: Proceedings of the Thirteenth Language Resources and Evaluation Conference, pp 258–266

Basile V, Cabrio E, Villata S, Frasson C, Gandon F (2016) A Pragma-Semantic Analysis of the Emotion/Sentiment Relation in Debates. In: 4th International Workshop on Artificial Intelligence and Cognition, New York, United States, Proceedings of the 4th International Workshop on Artificial Intelligence and Cognition

Bednarek M (2008) Emotion talk across corpora. Springer

Boucher J, Osgood CE (1969) The Pollyanna hypothesis. Journal of verbal learning and verbal behavior 8(1):1–8

Buechel S, Hahn U (2017) EmoBank: Studying the impact of annotation perspective and representation format on dimensional emotion analysis. In: Proceedings of the 15th Conference of the European Chapter of the Association for Computational Linguistics: Volume 2, Short Papers, Association for Computational Linguistics, pp 578–585

Cabrio E, Villata S (2018) The SEEMPAD dataset for emphatic and persuasive argumentation. In: Cabrio E, Mazzei A, Tamburini F (eds) Proceedings of the Fifth Italian Conference on Computational Linguistics (CLiC-it 2018), Torino, Italy, December 10-12, 2018, CEUR-WS.org, CEUR Workshop Proceedings, vol 2253

Da San Martino G, Yu S, Barrón-Cedeño A, Petrov R, Nakov P (2019) Fine-Grained Analysis of Propaganda in News Article. In: Proceedings of the 2019 Conference on Empirical Methods in Natural Language Processing and the 9th International Joint Conference on Natural Language Processing (EMNLP-IJCNLP), Association for Computational Linguistics, Stroudsburg, PA, USA, pp 5635–5645, https://doi.org/10.18653/v1/D19-1565

Gajewska E, Konat B (2024) Influencing hearts: Emotional appeals in public discussions on social issues on twitter, submitted

Gajewska E, Budzynska K, Konat B, Koszowy M, Kiljan K, Uberna M, Zhang H (2024) Ethos and pathos in online group discussions: Corpora for polarisation issues in social media. arXiv preprint arXiv:240404889 2404.04889

Hartmann J (2022) Emotion English distilroberta-base. https://huggingface.co/j-hartmann/emotion-english-distilroberta-base/

Hatfield E, Cacioppo JT, Rapson RL (1993) Emotional Contagion. Studies in Emotion and Social Interaction, Cambridge University Press, New York, https://doi.org/10.1017/CBO9781139174138, online publication date: June 2012

Herrando C, Constantinides E (2021) Emotional contagion: A brief overview and future directions. Frontiers in Psychology 12, https://doi.org/10.3389/fpsyg.2021.712606, section Emotion Science

Kocoń J, Miłkowski P, Kanclerz K (2021) Multiemo: Multilingual, multilevel, multidomain sentiment analysis corpus of consumer reviews. In: International Conference on Computational Science, Springer, pp 297–312

Konat B, Gajewska E, Rossa W (2024) Pathos in natural language argumentation: Emotional appeals and reactions. Argumentation 38:369–403, https://doi.org/10.1007/s10503-024-09631-2

Koszowy M, Budzynska K, Pereira-Fariña M, Duthie R (2022) From theory of rhetoric to the practice of language use: The case of appeals to ethos elements. Argumentation pp 1–27

Lawrence J, Bex F, Reed C, Snaith M (2012) AIFdb: Infrastructure for the argument web. In: Verheij B, Szeider S, Woltran S (eds) Computational Models of Argument, Frontiers in Artificial Intelligence and Applications, vol 245, IOS Press, pp 515–516, https://doi.org/10.3233/978-1-61499-111-3-515

Lukin S, Anand P, Walker M, Whittaker S (2017) Argument strength is in the eye of the beholder: Audience effects in persuasion. In: Proceedings of the 15th Conference of the European Chapter of the Association for Computational Linguistics: Volume 1, Long Papers, Association for Computational Linguistics, pp 742–753

Mohammad SM, Turney PD (2013) Crowdsourcing a word–emotion association lexicon. Computational intelligence 29(3):436–465

Nakov P, Ritter A, Rosenthal S, Sebastiani F, Stoyanov V (2019) Semeval-2016 task 4: Sentiment analysis in Twitter. arXiv preprint arXiv:191201973

Nguyen DQ, Vu T, Nguyen AT (2020) Bertweet: A pre-trained language model for English tweets. In: Proceedings of the 2020 Conference on Empirical Methods in Natural Language Processing: System Demonstrations, pp 9–14

Papacharissi Z (2015) Affective Publics: Sentiment, Technology, and Politics. Oxford University Press, New York

Stede M (2020) Automatic argumentation mining and the role of stance and sentiment. Journal of Argumentation in Context 9(1):19–41

Tajfel H, Turner JC (1986) The social identity theory of intergroup behavior. In: Worchel S, Austin WG (eds) Psychology of Intergroup Relations, Nelson-Hall, Chicago, pp 7–24

Visser J, Konat B, Duthie R, Koszowy M, Budzynska K, Reed C (2020) Argumentation in the 2016 US presidential elections: annotated corpora of television debates and social media reaction. Language Resources and Evaluation 54(1):123–154

Walton D (1992) The Place of Emotion in Argument. The Pennsylvania State University Press, University Park, PA

Walton D, Reed C, Macagno F (2008) Argumentation Schemes. Cambridge University Press, Cambridge, https://doi.org/10.1017/CBO9780511802034

Wierzba M, Riegel M, Wypych M, Jednoróg K, Turnau P, Grabowska A, Marchewka A (2015) Basic Emotions in the Nencki Affective Word List (NAWL BE): New Method of Classifying Emotional Stimuli. PLOS ONE 10(7):e0132305

Wierzba M, Riegel M, Kocoń J, Miłkowski P, Janz A, Klessa K, Juszczyk K, Konat B, Grimling D, Piasecki M, et al. (2021) Emotion norms for 6000 Polish word meanings with a direct mapping to the Polish wordnet. Behavior Research Methods pp 1–16

Yoosuf S, Yang Y (2019) Fine-Grained Propaganda Detection with Fine-Tuned BERT. In: Proceedings of the Second Workshop on Natural Language Processing for Internet Freedom: Censorship, Disinformation, and Propaganda, Association for Computational Linguistics (ACL), pp 87–91, https://doi.org/10.18653/v1/d19-5011, https://github.com/huggingface/

Younis R, Fernandes DdO, Gygax P, Koszowy M, Oswald S (2023) Rephrasing is not arguing, but it is still persuasive: An experimental approach to perlocutionary effects of rephrase. Journal of Pragmatics 210:12–23, https://doi.org/10.1016/j.pragma.2023.03.010

Chapter 9
Large Language Models for Pathos Mining

Abstract This chapter explores the application of Large Language Models (LLMs) for identifying emotional appeals (pathos) in natural language argumentation. Building on the interactional model of pathos presented earlier in the book (MIPA), it demonstrates how LLMs can support scalable annotation of rhetorical emotional appeals across languages and genres. The first study applies GPT-4 to competitive debate transcripts, analyzing categorical emotions (anger, fear, joy, sadness, disgust) and comparing results with manual annotation. The second study investigates multilingual political discourse in six European countries during pre-election debates, focusing on the prevalence of fear appeals in regions near armed conflict. In both cases, LLMs demonstrate fair intersubjective agreement with human annotation, though limitations remain in context sensitivity and mixed-emotion detection. The chapter argues that, despite their computational nature, LLMs can emulate annotation tasks when equipped with carefully designed prompts and evaluation metrics. Methodological reflections emphasize the need for hybrid approaches that combine computational and qualitative insights. The study confirms that pathos can be systematically mined using generative AI, offering a new avenue for interdisciplinary research in the areas of argumentation, computational linguistics, and emotion studies.

Keywords Pathos mining · Large language models · Emotional appeals in argumentation · Computational rhetoric · Multilingual political discourse · Fear appeal

9.1 Methodology for Pathos Mining with LLM

In Chap. 7, I presented an overview of how the studies in current argumentation mining, aimed at extracting argument structures from natural language dialogical interactions, are moving from classification-based approaches to generative approaches—using LLMs. As presented in Part I of this book, my view of pathos, following Aristotle and modern pragmatics, is an interactional one, where speaker appeals to the emotions of the audience for rhetorical gain. I have presented various

B. Konat, *Emotional Appeals in Argumentation*, Argumentation Library 47,
https://doi.org/10.1007/978-3-032-06332-8_9

ways of observing how those interactions occur in the natural world—with manual annotation of argument schemes in Part I, with psycholinguistic methods in Part II. Finally, here, in earlier sections of Part III, I presented how simple computational methods can be used to scale up pathos analysis with the application of automated lexicon-based methods for capturing words eliciting emotions and with the use of sentiment analysis models to capture audience reactions. All these methods are useful, but they have one limitation: they are mostly language dependent (like lexicons), or restricted to performing one task, like sentiment analysis models. For extraction (or mining) of emotional appeals (pathos) from natural language interaction, we would need a more versatile tool, closer to working with human annotators. Thus, in this section, I present results of using generative AI, or Large Language Model, in place of a human annotator, for pathos mining from natural language texts.

To operationalize pathos mining in alignment with our manual annotation scheme (Gajewska et al. 2024), I designed a prompt structure that mirrors human decision making as closely as possible while imposing strict constraints on the interpretation range. The Python function shown in Listing 9.1 was used to generate input messages for the OpenAI GPT-4 application programming interface (API), which allows the model to assess whether a given utterance contains pathos.

In this context, a prompt refers to the structured set of inputs that define the task of the language model. Large Language Models (LLMs) like GPT-4 are generative by default and tend to produce fluent, elaborative text unless explicitly instructed otherwise. However, for annotation-like tasks, generation must be restricted to controlled classification-style output that is interpretable and replicable.

To achieve this, the function in Listing 9.1 constructs a prompt composed of three message components. The *system message* defines the overall task and constraints, instructing the model to act as an analytical tool designed to detect rhetorical intent to elicit one of five emotions: joy, fear, disgust, sadness, or anger. It explicitly prohibits the model from inferring emotions outside of this set or interpreting affective content not clearly grounded in rhetorical intent. The *user message* presents the input text to be analyzed and frames the task as a focused evaluation of potential emotional appeal. Finally, the assistant message specifies the expected output format, instructing the model to return only one of the listed labels. These categories correspond directly to the annotation labels used in the human-coded data and reflect the decision logic adopted in the annotation guidelines.

This constrained prompting approach is critical when using LLMs for annotation-like tasks. It minimizes undesired generative behavior, such as overinterpreting vague language or inventing emotional content—phenomena often described as hallucinations in LLM contexts. By restricting possible outputs and anchoring them to the annotation scheme used by human coders, the prompt aims at compatibility and comparability across human and machine annotations.

```
def create_messages(text):
    return [
        {
            "role": "system",
            "content": "You are an analytical tool designed to
                identify rhetorical pathos in text. Your task is
                to analyze if the speaker's intention is to
                elicit specific emotions for rhetorical purposes.
                 The only emotions to consider are joy, fear,
                disgust, sadness, and anger. If the text does not
                intend to elicit these emotions, or if the
                intention is ambiguous, respond with 'No pathos'
                or 'Unclear', respectively. Do not infer emotions
                 not explicitly stated in the text, and do not
                identify emotions outside of the five specified."
        },
        {
            "role": "user",
            "content": f"Analyze the following text for pathos: {
                text}"
        },
        {
            "role": "assistant",
            "content": "Based on the analysis, is there an
                attempt to elicit an emotion? If so, identify
                only one of the specified emotions: joy, fear,
                disgust, sadness, anger. If not, respond with 'No
                 pathos'. If uncertain, respond with 'Unclear'.
                Do not identify emotions not listed."
        }
    ]
```

Listing 9.1 Prompt used for OpenAI GPT-4 API

The function was integrated into a broader workflow that processed natural language argumentation in the two studies presented below. Each sentence of the text was passed through the API using this prompt structure, and the output was saved for further quantitative and qualitative analysis.

Limitations of Large Language Models, just like with any deep neural network, include their non-deterministic nature: they can return slightly different results each time, and different generations of models will return different results over time. It is akin to the non-deterministic nature of human annotators, whom we also cannot fully control and who will make their decisions differently as they are influenced by a myriad of factors. Just like for humans, for LLMs we cannot make sure that they will become deterministic machines, but we can attempt at consistency by applying measurements of consistent annotations and improve annotation schemes transferred into prompt. For the study presented here, I repeated the analysis on a different model (GPT-3.5 turbo) on a different date, and observed very similar results with differences not larger than 5%, which resulted in Cohen's Kappa of 0.4. This is only a fair level of agreement, and of course we would like to achieve more, however, the level of agreement between human annotators reported in Gajewska et al. (2024) was of a similar level (with slight differences in calculation methodology). To further test the stability of GPT-4 annotations, I performed additional automated analysis on sub-samples of the human annotated corpus reported in Gajewska et al. (2024), again, obtaining the category-wise Cohen's Kappa levels between 0.35 and 0.45, indicating fair agreement (with observed percentage agreement levels between 0.52 and 0.31), showing not high but stable values across different corpora and genres. I also calculated the weighted F1 score with automated annotation measured against human annotation. The results are similar (F1 = 0.56), and future methodological considerations should decide whether we would treat LLM as just another annotator (and therefore calculate Cohen's Kappa), or we would use human data as a gold standard (thus using the F score as a metric). Taking all this into account, we can see that the perception of emotional appeal is not fully subjective and can be communicated intersubjectively both between human annotators and between humans and machines, yet there is still a way for the development and improvement of the techniques for pathos mining.

For both source materials here, competitive debating and pre-election debates, I assume that the context guarantees persuasive use of emotional appeals. Therefore, I will not present entire argumentative structures, assuming that in both cases speakers are in the role of persuading audience and each emotional appeal automatically fulfills the condition of being used for persuasive gain, as drafted in the interactive pathos model in Part I.

I would like to conclude this section with two methodological notes. First, the most effective methodological advances can be achieved through a combination of methods, as each offers unique strengths and limitations. Emotional lexicons are grounded in psychological theories of emotions, and their creation is best done in controlled environments. Human annotators can be trained and re-trained using detailed annotation manuals, ensuring a high-quality transfer from argumentation theory to language analysis practice. Classification models are valuable due to their simplicity, availability in multiple languages, and ability to be fine-tuned for specific tasks. Large Language Models (LLMs) in this context present both advantages and limitations. Their most remarkable feature is their versatility: they can be used for a wide range of tasks without requiring pre-training. Similarly to human annotators, they rely on clear instructions that can be detailed and include references to the theoretical background informing those instructions. However, an LLM is still just a model, a deep neural network designed and trained to emulate human verbal behavior as closely as possible. Although it is certain that they do not experience emotions in the human or animal sense, they are designed to react to language and produce outputs that closely mimic human responses. Therefore, while caution is warranted in entrusting LLMs with a variety of serious tasks, they may still offer reliable recognition of emotional appeals as realized in language.

The second methodological note recapitulates earlier points: Large Language Models (LLMs), such as GPT, do not feel emotions—they process language based on patterns in the data on which they have been trained. In an earlier generation of sentiment analysis models discussed in the last chapter, which were capable of identifying positive or negative sentiments in texts, it was never claimed that these models experienced emotions. The same applies to LLMs, even though they are significantly better at understanding context and generating human-like responses. LLMs recognize emotions in text and produce emotional language not because they feel emotions but because they have been trained on extensive datasets, including linguistic expressions and emotional appeals. This process can be illustrated with the metaphor of a thermostat, which can detect temperature and adjust it without "feeling" hot or cold. Similarly, LLMs can generate and recognize emotional language without experiencing emotions, relying solely on computational processes rather than thought, self-awareness, or feelings.

Incorporating these two methodological observations allows us to see how LLMs can be utilized in tasks focused on identifying pathos in natural language. By combining their computational capabilities with methods grounded in psychological theories and human expertise, we can create a useful framework for pathos mining.

9.2 AI Applications of MIPA in Competitive Debating

9.2.1 Quantitative Results in Competitive Debating

As a material for this study I use natural language argumentation from competitive debates performed by students (Obrębska et al. 2025). The details of this experiment are described in Chap. 6, here let us just remind that we engaged two teams of students in competitive debate (after a training session with professional debaters) and there were two debates, each with its own topic: 1. "We should stop using social media". 2. "We should decide on career path based on money, not passion". LLM annotation of this material was performed specifically for this book. Below I present results of automated annotation with Large Language Model—Chat GPT-4 completion API (15.11.24) with the prompt presented in previous section. What is important, as can be seen in the prompt, I did not ask the model for justification of its decision, therefore all interpretations presented below are mine, not based on automated justification.

As we can see in the pie chart with the general distribution of emotional appeals in competitive debate transcripts, the dominant category (75.7%) remains "No pathos". This means that the model used for annotation decided that there was no intention of eliciting emotions in the audience in these cases. Emotional appeal was identified in 21.2% cases, and 3.1% was marked as unclear or returned incorrect answer. An incorrect answer means in this context that it did not contain any emotion name.

Figure 9.1 presents these 21.2% cases in which pathos was identified split by category of emotion, also presented in detail in Table 9.1. As we can see, the most dominant category the speakers appeal to is fear, followed closely by sadness and happiness. This is really interesting—as a reminder, these are students, speaking to the audience consisting of their peers, however, with the clear goal of winning the debate and being as competitive as possible. In this context, it looks that next to

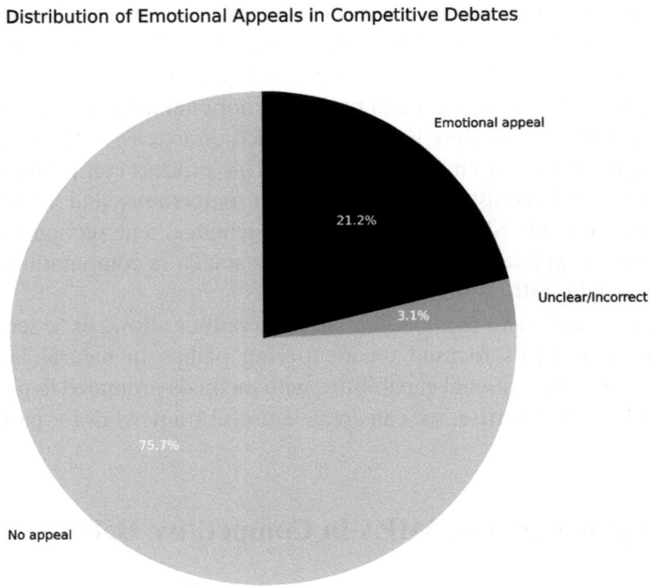

Fig. 9.1 Pathos in competitive debate summary

Table 9.1 Summary of sentence categories and their distributions (Total: 552 sentences)

Category	Frequency	Percentage
Appeal to Anger	5	1
Appeal to Disgust	4	1
Appeal to Fear	48	9
Appeal to Joy	26	5
Appeal to Sadness	34	6
No Pathos	418	76
Unclear/Incorrect	17	3
SUM	522	100%

appealing to happiness, which could be a natural choice in a relaxed environment of students' discussions, they decided to use fear appeal, possibly perceiving it as most effective and strong. In the next sections, we will observe examples of each categorical emotional appeal and how it was realized by the participants (Fig. 9.2).

9.2.2 *Case Studies in Competitive Debating*

9.2.2.1 Anger

After the distribution of emotional appeals identified through pathos mining, we now turn to a qualitative examination of these emotions in context. This section

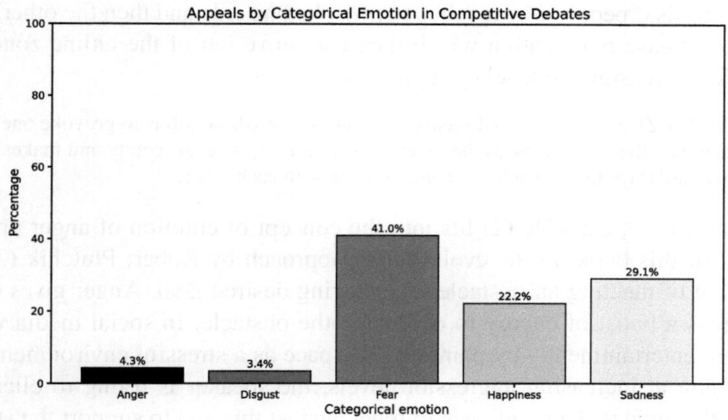

Fig. 9.2 Pathos in competitive debate—by emotion category

presents selected examples that inspect how Large Language Models capture and categorize emotional appeals in real argumentative speech.

In our data, the pathos mining process identified five instances of appeals to anger. In example (1), the speaker argues for the claim "We should stop using social media", arguing that the use of social media inevitably results in its control.

(1) Speaker VU8C: In my opinion, this is a really bad thing because it allows social media to control us.[1]

In example (1), the speaker seeks to elicit anger by highlighting that social media control us—a situation that most individuals resist. This pathos-based strategy emphasizes the undesirable nature of being controlled. The speaker also incorporates the appraisal component of emotion, expressing their opinion that this control is negative. This example illustrates how emotional appeals in persuasive speech often intertwine with the speaker's expression of personal emotions and stances.

This observation also connects to Ekman's concept of emotion families, discussed alongside the concept of appraisal in Part II of this book. The sentence may also involve an appeal to fear, but the appeal to anger is more prominent. This highlights the complexity of annotating pathos in natural language argumentation, presenting challenges for both human annotators (trained or untrained) and automated language models. As this book explores intersubjective methods for pathos analysis that support operationalization within the proposed model, this example shows both the potential and the limitations of such methods.

Another example of appeal to anger is (2), where the speaker is also arguing for the case of not using social media, by showing how professionally organized

[1] Polish original text [Pol.]: *To jest moim zdaniem bardzo zła rzecz, ponieważ przez to media społecznościowe nas kontrolują.*

groups of trolls ("people whose job is to provoke one side and then the other") work online to increase polarization which then can move out of the online zone to the real world, increasing the level of aggression.

(2) Speaker ZUZ9: So on social media, you get people whose job is to provoke one side and then the other, which raises the overall level of aggression in society and makes it harder and harder for the two sides to communicate with each other.[2]

As we can see, example (2) fits into the concept of emotion of anger presented in Part II of this book, in the evolutionary approach by Robert Plutchik (2003)—as a feeling of meeting an obstacle to achieving desired goal. Anger gives humans and animals a boost of energy to overcome the obstacle. In social media we seek relaxation, entertainment—by painting this space as a stressful environment full of trolls aiming at increasing aggression levels, the speaker is trying to elicit anger which will convince their audience to rebel against this and to support the stance of resignation from the use of social media.

While in the example (2) the argument itself is compelling and it surely relates to the emotional side of the listeners, is it surely an appeal to anger? Does the speaker intentionally try to elicit anger in hearers for rhetorical gain, in accordance with the conceptualization of pathos provided in Part I of this book? It is not that clear in this case, we know that definitely example (2) talks about anger and aggression, however, does it elicit one more or less universally?

In example (3) the situation is slightly different, as on the linguistic surface we might perceive that this fragment relates more to eliciting or reminiscing the emotion of fear ("we are stressed") rather than anger. The speaker is arguing for the thesis claiming that we should decide on career path based on salary rather than passion, by showing how in everyday work this pleasure and passion can be thwarted by the fact that even in the work that we are passionate about we can be forced to perform actions which are maybe not so pleasant. In the earlier part of this excerpt, the speaker uses the case of acting—a profession mostly selected with passion and a dream of playing important, deep characters. The speaker then presents the situation in which an actor, actress is asked to play a tree in a play. This vision of passionate artists feeling forced is meant to elicit anger, the feeling of defiance against an obstacle.

(3) Speaker 8WDX: We are stressed by the fact that we want to do something, dedicate our lives to something we love, and then someone tells us to do something else.[3]

In example (5), the speaker describes the emotional impact of pursuing a career driven by passion with a focus on burnout and dissatisfaction. The speaker uses the case of graphic designers, illustrating how creative professions often involve similar

---

[2] [Pol.] *Więc na social mediach pojawia się wiele osób, które mają za zadanie zdenerwować jeden obóz i drugi obóz, co ogólnie podnosi poziom agresji w całym społeczeństwie i sprawia, że coraz trudniej się dogadać pomiędzy jednym a drugim obozem.*

[3] [Pol.] *Jesteśmy w takim razie zestresowani faktem, że chcemy coś robić, poświęcamy nasze życie na coś, co uwielbiamy, a ktoś nam każe robić coś innego.*

challenges. The speaker decides to use emotion names to appeal to the audience, contrasting "love" and "hate", which we can interpret in terms of "emotion names" of the French tradition, as presented in Part II of this book.

(4) Speaker 8WDX: At this point, passion is what leads me to burnout because I've started to hate that something I once loved has turned me into someone fulfilling other people's tasks. The same thing happens to graphic designers.[4]

In summary, in the context of anger appeals, we still should remember that LLMs are still just language models, despite their sometimes remarkable and human-like capabilities. In examples (2) and (5) we can observe that the identification of appeal to anger could be motivated by the presence of the word "hate". While we can argue that these cases are really appeals to anger, still, LLMs can easily mix up the perspective of expressed emotions with pathos and be overly suggestive to strong emotion names appealing on the linguistic surface, rather than focusing on the intention of the speaker.

9.2.2.2 Disgust

The cases of appeals to disgust are somewhat less clear and straightforward (including more "mixed" cases) than the cases of appeals to anger previously discussed. Example (5) paints the picture that without a doubt would elicit disgust in most listeners: a job that takes place in a landfill, involving fighting 14 rats a day. However, it is obvious that this appeal is not used here in a straightforward manner—the speaker paints a picture that is somehow humorous in nature, bringing up cartoon-like vision of landfill work. The speaker combines these contrasting emotions, disgust and humor, to engage the audience better and to enhance their memory of this argument.

(5) Speaker 8WDX2: If you're passionate, then this job could even be at a landfill, where you'd have to fight off 14 rats every day.[5]

Another speaker, in example (6) argues for the use of social media, by attempting to elicit disgust toward older, more outdated—in their opinion—form of communication, which is an email. The speaker uses the phrase "people just don't want to look at it" which in the original conveys the feeling of reluctance, lack of motivation, or disinterest. It is not strictly disgust, as defined in Part II in terms of encountering poisonous stimuli, resulting in the need to eject it from the system. This physical understanding of the emotion of disgust does not fit in here. However, Robert Plutchik (2001), in his circumplex model of emotions (often dubbed the "emotion

[4] [Pol.] *W tym momencie to pasja prowadzi mnie do wypalenia, ponieważ zaczęłam nienawidzić tego, że coś, co kiedyś kochałam, zmieniło mnie w kogoś, kto wypełnia zadania innych ludzi. To samo dzieje się z grafikami.*

[5] [Pol.] *Jak masz pasję, to ta praca może być nawet na śmietnisku, gdzie tam trzeba się bić z 14 szczurami dziennie.*

wheel") presents a milder, less intense form of disgust—boredom. In the context of this example, the speaker does not attempt at eliciting physical disgust (as the one in previous example, resulting from the view of the dumpster and rats), but instead uses the concept of something outdated, boring, and ineffective to elicit milder version of disgust (boredom) towards the e-mail as an alternative to social media.

(6) Speaker UFYQ: The mail system is not updated very often, so it is outdated. People don't want to look at it, they're checking their mail less and less, and that's why they're missing out on important information.[6]

The mixed nature of disgust cases is best illustrated in example (7), where the speaker paints the picture of passion towards the job, turning into hate over time, making an argument against choosing a profession based on passion. Although we might intuitively perceive the vision of a worker starting with passion and gradually slipping into—previously mentioned—boredom and effectively into disgust towards the job they once loved, what is explicitly mentioned in the example is "hate". Again, the wheel of emotions (Plutchik 2001) proves useful in this context, as it shows anger and disgust as two neighboring emotions that, when mixed, can lead to the emergence of a dyad emotion of contempt. Similarly, the concept of the "anger family" (Plutchik 2003) also covers mixed emotions of wanting to fight and eject the stimuli. This mixture best describes the feeling the speaker is trying to elicit in listeners here, trying to make the audience feel the emotion of boredom, anger, and sympathizing with the contempt of a person who first was passionate over their job, but now cannot stand thinking about it.

(7) Speaker KGDD: And something that was once their passion turns into something they hate.[7]

All the cases described in this section illustrate well the complex nature of emotional appeal to disgust. One determining factor here is the topic of the debates (which is arbitrarily assigned to the speakers)—the topics were abstract, related to non-physical environment strongly entrenched in technological civilization. It would seem difficult to appeal to the emotion related to "ejecting poison" (Plutchik 2003). However, evolutionary theory of emotions provides us with contexts and shows how less intense versions of disgust, i.e. boredom, or appeals to disgust mixed with anger, can be effectively used to add pathos to argumentation even related to abstract domains.

[6] [Pol.] *Poczta nie jest tak często aktualizowana, przez co jest staromodna, ludziom nie chce się na to patrzeć, coraz rzadziej zaglądają właśnie na pocztę i przez to omijają ważne informacje.*

[7] [Pol.] *I coś, co było na początku ich pasją, staje się znienawidzonym czymś.*

9.2.2.3 Fear

Appeals to fear always had special place in argumentation theory. From the concept of *ad baculum*, a simple threat used to force the listener to obey, to the modern version of fear appeal argument scheme proposed by Walton (2008). Walton had even devoted a separate book to threats and fear appeals (2013), emphasizing a special place of this type of pathos among other emotional appeals. Fear is a clear, strong emotion, leading to fast physical reactions, and thus an effective elicitation of fear in audiences can be a powerful persuasive tool.

Example (8) was already present in the Introduction of this book, and here, using the concept of categorical emotions and LLM classification, we can analyze it in more concrete terms. This case shows how the topic of social media, often perceived as harmless, can be reframed to elicit fear. The speaker argues that social media use can lead to addictions that, while different from those associated with substances like alcohol or cigarettes, can have significant psychological effects. The speaker emphasizes that even though the physical effects may not be immediately visible, the psychological impact can be severe, making it easy to develop such an addiction. We can observe that by introducing comparison to the well-known harmful additions, the speaker is attempting at increasing their fear appeal to make their argument against social media use.

(8) Speaker GPAD1: Social media leads to addictions; these are very simple addictions and often not taken as seriously as, for example, alcohol or cigarettes. Even if you don't see the physical effects, it doesn't mean they're not there. The psychological effects can be really devastating, and it's super easy to fall into this kind of addiction.[8]

Surprisingly, the speakers arguing for the opposite side, also managed to create fear appeals showing unwanted consequences of not using social media. In example (9) the speaker uses the contrast of painting something positive, something that would definitely appeal to the group of young, ambitious students (personal development) and contrasts it with the vision in which the lack of social media use can lower this positive process, thus arguing for the use of social media. This surprising, but smart strategy aligns with Plutchik conceptualization of fear—we have a cherished object and we do not want to lose it. Fear is a threat of losing something of value. The speaker in this example is using this principle to elicit fear in the audience and obtain rhetorical gain.

(9) Speaker 5EF9: Firstly, the lack of social media use can actually lower personal development.[9]

---

[8] [Pol.] *Media społecznościowe prowadzą do uzależnień, są to bardzo proste uzależnienia i często nie brane pod uwagę tak poważnie, jak na przykład jakiś alkohol czy papierosy. Choć nie widać może efektów fizycznych, co nie znaczy, że ich nie ma, psychiczne na pewno mogą być też druzgocące i bardzo łatwo popaść w takie uzależnienie.*

[9] [Pol.] *Po pierwsze, że brak używania social mediów może spowodować obniżenie samorozwoju.*

Similar strategy of eliciting fear of unwanted consequences in the form of "certain behavioral issues" is used in example (10) where speaker argues against choosing career path based on passion. As we can see, regardless of the topic of the debate or the stance (for or against the discussed issue), the speakers choose and performed this strategy in the same way.

(10) Speaker KCNU: Moreover, working and pursuing a passion in the same place can lead to certain behavioral issues.[10]

While previous examples of fear appeal painted a vague picture of unwanted consequences, the speaker in example (11) decides for more gruesome details to increase the power of their argumentation against choosing the job we are not passionate about.

(11) Speaker EC25: So, when it comes to the long-term consequences of doing a job that doesn't interest or inspire us, it's really about stress. And of course, we know that stress has a negative impact on our entire life; it can lead to a shorter lifespan, a higher risk of diseases, and all those negative consequences in personal life, even affecting the people around us.[11]

According to the speaker in example (11), when we perform a job "that doesn't interest or inspire us", the long-term consequence would be constant stress. The speaker follows with further consequences including diseases, certain unspecified "negative consequences", but also death (even though the speaker decides against using this word and uses euphemism of "shorter lifespan" instead, the threat is there). To drive the point home, the speaker decides to not only list possible morbid consequences for the worker themselves, but also for the family and close ones, possibly hoping for increasing the level of fear appeal. Is it effective? In this case we can intuitively perceive that the speaker is reaching close to the level of overdoing fear appeal, of using visions too strong and too far away in time for them to be convincing. Perhaps the speaker can be accused here of slippery slope fallacy—of course they cannot argue that selecting job which we are not passionate about will lead to premature death. This is why the speaker here moves to euphemistic form, however still, these very strong consequences lead to the risk of fear appeal not being effective. However, the vision of stress, disease, and death is already put in the minds of the audience.

Another strategy of avoiding the trap of too strong appeal or slippery slope fallacy, next to the euphemisms which move away accusation of using too strong language, is the use of rhetorical questions. Rhetorical questions or assertive

[10] [Pol.] *Co więcej, pracowanie i uprawianie pasji w jednym miejscu może spowodować pewne problemy behawioralne.*

[11] [Pol.] *To właśnie jeżeli chodzi o takie długofalowe konsekwencje wykonywania zawodu, który nas nie interesuje, nas nie pasjonuje, no to właśnie jest ten stres. No i wiadomo, że stres wpływa negatywnie na całe nasze życie, no bo może doprowadzić właśnie do skróconej długości życia, większego ryzyko chorób, właśnie te negatywne konsekwencje w życiu osobistym, gdzieś tam na osoby z naszego otoczenia.*

questions allow speakers to seemingly move away from owning the statements they put into the dialogue, but still the assertion is put into the discussion. Let us observe example (12), where the speakers appeals to one of the strongest negative tropes—the harm of a child. The speaker leaves this unsaid, suspending it in a rhetorical question. However, the audience already knows, already senses the unease, that something bad is likely to happen to the child. After all, the speaker wouldn't leave such an implication hanging if it were something positive. At the same time, the speaker preserves face in case of accusations, as there is no explicit claim or evidence that something bad will happen to the child.

(12) Speaker KGDD: If a father is fully devoted to his passion and that becomes the main factor in choosing his job, then what will happen to his child?[12]

We can see how appeals to fear are more uniform than other categories of appeals, as the speakers mainly use the same strategy of showing very serious future consequences, mostly related to either physical or mental health of either the potential audience or the close ones. And they will paint a very bleak picture of serious consequences that will occur if the audience does not adhere to the arguments proposed by the speakers.

9.2.2.4 Joy

Speakers also argued for their cases with appeals to joy, mostly by showing possible future positive consequences of actions. In case of example (13), the speaker infuses their claim for the stop of social media usage with numerous positive words: grassroots movement, positive impact, development and well-being. The speaker puts it all in the frame of the future benefits, arguing for the current issue.

(13) Speaker GPAD2: We want to show that a grassroots movement to stop using social media will positively impact our development and our well-being.[13]

When arguing for the selection of career based on passion, the speaker in example (14) is also attempting at elicitation of the emotion of happiness in hearers, by presenting a vision of feeling happy after performing work which we like.

(14) Speaker P2X2: I would also like to mention that if we like the work we do, we are happier.[14]

---

[12] [Pol.] *Jeśli ojciec będzie się poświęcał pasji i to będzie jego najważniejszą kwestią w wyborze pracy, no to co się stanie z dzieckiem?*

[13] [Pol.] *Chcemy pokazać, że zmiana oddolna, żeby przestać korzystać z tych mediów społecznościowych, wpłynie pozytywnie na nasz rozwój, na nasze samopoczucie.*

[14] [Pol.] *I chciałabym tutaj wspomnieć również, iż jeżeli lubimy pracę, którą wykonujemy, to jesteśmy szczęśliwsi.*

In example (15), the speaker uses slightly different approach to argue for the same issue—instead of focusing on the feeling of personal happiness, they refer to the objective indicators of job well-done: high quality and large clientele. The speaker claims that choosing the job based on passion would lead to higher quality, and a secure future in this profession. Again, this is an appeal to joy, but in a slightly different manner, possibly closer to serenity (or calmness) which in Plutchik's wheel of emotion is treated as a lighter version of joy.

(15) Speaker P2X2: It means that if I like what I do, I do it more thoroughly, it has higher quality, so I will really have clients, and I'll have a certain kind of future secured.[15]

9.2.2.5 Sadness

When appealing to the emotion of sadness, speakers often employ constructions closely resembling the operationalization of this emotion as the loss of a valued object, as described by Ekman and Plutchik. Example (16) illustrates this perfectly, as the speaker evokes the narration of a lost friend, with whom we were once connected by a shared passion. The speaker directly talks about "emotional loss" the hearer would feel in such scenario, thus using an appeal to sadness to support their case for choosing job based on passion rather than money.

(16) Speaker KCNU: For example, your friend no longer pursues that passion, the one that really connected you both. You think about the emotional loss you feel because that friend is no longer around.[16]

Another example—(17)—similarly tells a story of professional regret, where the speaker appeals to sadness by highlighting emotional loss. It describes a person who chose a stable but unfulfilling career in banking, thinking the financial benefits would outweigh the drawbacks. Over time, the psychological strain of doing work they disliked grew. They found themselves unable to pursue their dream career, such as becoming an actor. Here, the "loss object" is the unattainable dream job, representing the lasting emotional impact of sacrificing passion for financial security.

(17) a. Speaker EC25: And an example might be a guy who went into banking because he thought the good money would be worth it if he endured a little hardship. At first, maybe it wasn't so bad, but over time he began to feel the negative psychological effects of doing a job he didn't enjoy.

[15] [Pol.] *Oznacza to, że jeśli lubię to, co robię, to robię to bardziej solidnie, ma to większą naprawdę będę miała tych klientów i będę już miała zapewnioną jakąś taką przyszłość.*

[16] [Pol.] *Na przykład twój znajomy już nie uprawia tej pasji, z którą jakby dużo was łączyło. Myślisz o tym, jaką masz stratę emocjonalną przez to, że tego znajomego nie ma.*

b. And he couldn't switch to his dream job, like becoming an actor, because he no longer had a chance to make the right contacts or integrate into that specific environment, one that would allow him to succeed and earn enough to support himself.[17]

9.2.2.6 Conclusions and Limitations

While the use of LLMs for pathos mining can be an effective tool, it has its limitations. The aim of this section is to illustrate those with selected examples. First, the main problem lies in the scarcity of pathos appels in most datasets. A dominant category in all recognized results is still "no pathos". As we have seen in the prompt at the beginning, the machine was instructed, just as student annotator would be, to only annotate when it was sure that there was this appeal to emotion. Let us analyze a few cases from the category "no pathos". In the example (18) we have a case of rich in content, seemingly factual argument for the use of social media. The speaker is listing useful content types which can be found in social media. The use of words ("interesting", "refined") indicates that the speaker has positive stance towards these contents, however LLM marked this as not containing any pathos appeals. We might assume that this is a borderline case, with certain expression of positive emotions, but without clear intention of elicit one in the audience for rhetorical gain.

(18) Speaker UFYQ2: What's more, there is also interesting information there, different facts about space, various fascinating facts, kind of refined knowledge.[18]

Another borderline cases include sentences where potentially the speaker would address something that can be viewed as positive in terms of universally accepted positive values, such as human rights, as in example (19).

(19) Speaker 5EF9: I also wanted to mention that, thanks to social media, we are able to better protect human rights.[19]

[17] [Pol.] *No i takim przykładem może być taki pan z bankowości, który na początku poszedł tam pracować, no bo by mu się wydawało, że dobre pieniądze, trochę się przemęczy i na początku może nie było tak źle, ale z biegiem czasu po prostu zaczął coraz bardziej odczuwać te negatywne konsekwencje psychiczne, jeżeli chodzi o wykonywanie pracy, się nie podoba. No i nie mógł się już przebranżowić na swoją wymarzoną pracę, na przykład już nie miał szansy, żeby nabyć odpowiednich kontaktów i żeby jakoś tak się wdrożyć w to specyficzne środowisko, które umożliwiłoby mu osiąganie sukcesów i zarabianie na takim poziomie, żeby móc się spokojnie utrzymać.*

[18] [Pol.] *Co więcej, są tam również ciekawe informacje, różne informacje o kosmosie, różne fakty, ciekawe informacje, takie wyszukane.*

[19] [Pol.] *Chciałam jeszcze również powiedzieć o tym, że dzięki social mediom jesteśmy w stanie lepiej zadbać o prawa człowieka.*

The annotation scheme included a category labeled "unclear." In the annotation results, the model categorized numerous questions (frequently rhetorical ones) as "unclear," as demonstrated in example (20):

(20) Speaker 5EF9: How do situations look right now where people are stepping away from social media?[20]

Cases like example (20) were categorized as "unclear" due to inherent ambiguities in their structure and function, which present challenges for both human and automated annotation of pathos in natural language argumentation. Language models, including GPT, rely heavily on explicit emotional markers to interpret intent, particularly when constrained to specific emotions. However, certain linguistic features, such as rhetorical questions, unfinished sentences, and statements combining stance with phatic expressions, complicate this task. Rhetorical questions, for instance, often aim to provoke reflection rather than to elicit a specific emotion, making it difficult to categorize the speaker's intent. Similarly, unfinished sentences lack the necessary context for clear interpretation, while mixed-function statements may serve multiple purposes: providing information, maintaining politeness, or prompting further dialogue.

One very interesting, erroneous example includes speaker P2X2 performing a complex strategy of contrasting positive and negative elements in one utterance (21).

(21) a. Speaker P2X2: I would also like to mention here that if we like the work we do, we are happier.[21]
b. If there is work that we don't like and that costs us a lot, we come home, for example, and have to somehow unwind.[22]
c. Sometimes we need to unwind in a way, such as reaching for various substances that are supposed to make us feel happier.[23]
d. For example, we reach for alcohol, which means we have an increase in addictions.[24]
e. If we do something with passion, we don't need to cheer ourselves up because we are already satisfied.[25]

Opening sentence—example (21) (a) argues for choosing job based on passion rather than money by claiming that "if we like the work we do, we are happier". The model marked this as appealing to the emotion of joy, and I assume most readers

[20] [Pol.] *W jaki sposób wyglądają w tym momencie sytuacje, w których osoby odstawiają social media?*

[21] [Pol.] *I chciałabym tutaj wspomnieć również, iż jeżeli lubimy pracę, którą wykonujemy, to jesteśmy szczęśliwsi.*

[22] [Pol.] *Jeżeli jest praca, której nie lubimy, dużo nas kosztuje, wracamy sobie na przykład do domu i musimy jakoś odreagować.*

[23] [Pol.] *Czasami musimy odreagować w taki sposób, jak na przykład sięganie po różne substancje, które mają sprawić, że będziemy weselsi.*

[24] [Pol.] *Na przykład sięgamy po alkohol, czyli że mamy wzrost uzależnień.*

[25] [Pol.] *Jeżeli robimy coś z pasją, nie musimy sobie potrafiać humoru, ponieważ już jesteśmy zadowoleni.*

would agree with this interpretation. The second sentence here, (21) (b), presents a dull image of a person coming back from the job they do not like, and looking for a way to "unwind". This is classified as appealing to sadness, which may not be clear, however might be reasonable. Yet, the case (21) (c) is marked as appealing to joy, which is an obvious error. Yes, the speaker in this sentence mentions the emotion of happiness, however it is in the context of psychoactive substances use, and speaker's probable intention is to appeal to fear. Only in the context of the three previous sentences it becomes clear that the speaker here is attempting at creating a vision of an unwanted job leading to the development of substance abuse, which is a clear appeal to fear. This is deepened in the next sentence, (21) (d) where the speaker is making their bleak vision more precise, mentioning alcohol as such a substance, and connecting it to the increase of addictions in society. This sentence is erroneously marked as not appealing to pathos, as its whole appeal is apparent only in the context of previous sentences. The speaker finishes with a circular structure, returning to a positive note from the beginning—the sentence (21) (e) presents choosing the job based on passion as the solution to the problem. What does this example tell us about human strategies of pathos appeals and the possibilities of their LLM annotation? In example (21), the speaker employs an emotional seesaw, mixing positive and negative emotional appeals as rhetorical strategy. While easy recognizable by human analyst, these subtle differences seem to escape the attention of LLM. This is partially the artifact of adopted methodology—I decided to feed the model with one sentence at a time via API, to ensure precision in answers. However by doing so, I minimalized contextual window to just one sentence, cutting of all co-text before and after it. While LLMs are able to process large contextual windows, this contributes to their unpredictability. Other approaches in argument mining had already explored the role of context, as described in previous chapter, therefore I believe that including larger co-text for pathos mining in a systematic manner is still a promising research avenue.

To sum up, with several limitations, LLMs can be effectively employed for pathos mining. I believe this approach represents an exciting intersection of argumentation theory, cognitive science, and computational linguistics, with potential implications for both theoretical research and practical applications in education and technology. By continuing to refine these methods and address the challenges, we, as argumentation community, can contribute to a forward-looking, interdisciplinary field that has the potential to reshape how we understand and teach persuasive language in the digital age.

What is remarkable is that in this experimental setting, students intuitively employed a multitude of pathos strategies, addressing all categorical emotions. All analyzed examples were drawn from the transcripts of student competitive debates conducted as part of the experiment described in Part II of this book. Even though participants were unaware of the experiment's purpose or its connection to the study of emotional appeals, they effectively constructed a palette of appeals that almost completely cover Plutchik's wheel of emotion.

9.3 AI Applications of MIPA in Political Discourse

This section presents another study using pathos mining with LLM. Pathos mining in multi-language pre-election debates explores how a country's geopolitical context—particularly proximity to military conflict—might influence the distribution of emotional appeals during debates. The original study (Kiljan and Konat submitted), in collaboration with political scientists, quantitatively analyzed fear as a rhetorical strategy, finding its use more prevalent in countries closer to war. Building on those findings, this chapter incorporates a qualitative examination of diverse pathos cases, pointing to both cross-linguistic similarities and culturally specific patterns in the expression of emotional appeals.

In 2023, parliamentary election debates in six European countries—Estonia, Finland, Poland (bordering Russia) and Greece, the Netherlands, Spain (more distant from Russia)–all took place amidst the ongoing Russian aggression against Ukraine, which began in February 2022. This war, the largest military conflict in Europe since World War II, caused widespread destruction and raised significant concerns about European security. Countries near Russia, such as Estonia, Finland, and Poland, which have historically been wary of Russian expansionism, experienced heightened fears of potential aggression. These anxieties were reflected in their political discourse, with fear of Russian threats becoming a central narrative in shaping voter concerns.

For our corpus study, we automatically transcribed six debates (one from each country) using OpenAI Whisper. Each debate ranged from 1 to 3 hours in length, resulting in corpora of 800 to 2000 sentences per country. We then divided the dataset into two sub-corpora: "Near" countries (Estonia, Finland, Poland) with approximately 2000 sentences and "Far" countries (Greece, the Netherlands, Spain) with about 5000 sentences. These corpora were annotated for pathos categories using the prompt described in the previous section. To test our hypothesis, we compared the distributions of categorical emotions in appeals between the "Near" and "Far" countries.

9.3.1 *Quantitative Results in Political Discourse*

Summary of results presents as follows. Figure 9.3 shows that emotional appeals constitute 16.3% of all analyzed cases, with the majority (74.3%) lacking any explicit emotional appeal. A smaller proportion (9.3%) was classified as unclear or incorrect interpretations of pathos. We can then see that the proportion of emotional appeals is smaller than in the previously analyzed competitive debate corpus.

Figure 9.4 and Table 9.2 detail the distribution of emotional appeals by category. Among the identified emotions, appeals to anger (37.4%) and fear (31.6%) dominate, followed by sadness (18.2%), happiness (8.9%), and disgust (3.9%).

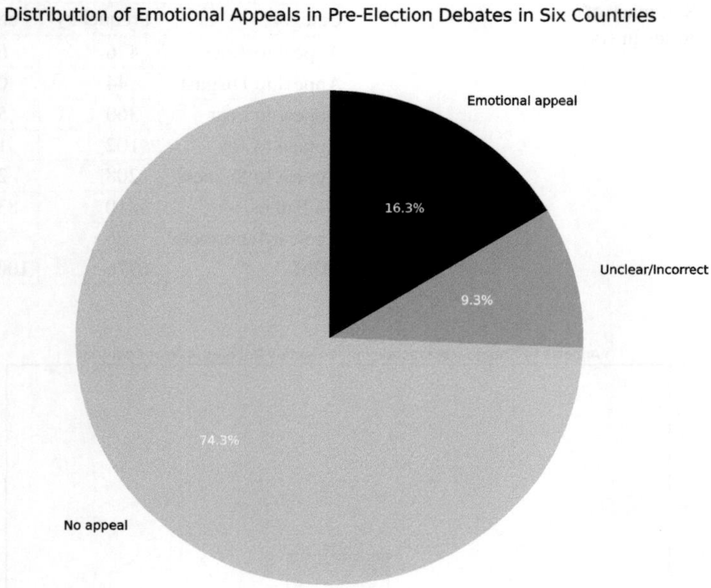

Fig. 9.3 Pathos in six countries summary

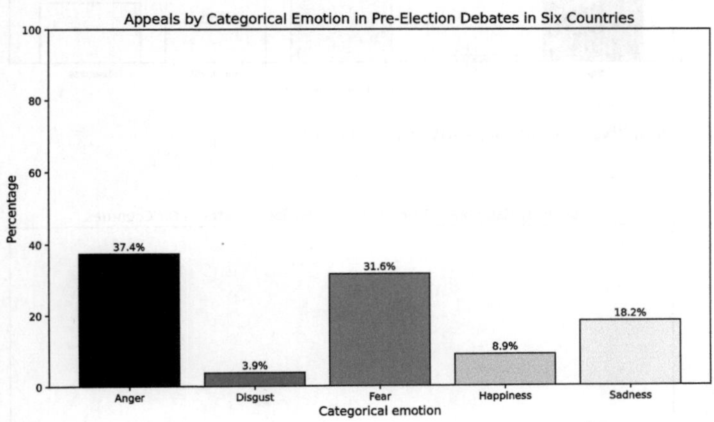

Fig. 9.4 Pathos in six countries—by emotion category

A comparison of countries closer to and farther from military conflicts shows a contrasts in emotional strategies. Figure 9.5 shows that in near countries, fear dominates the emotional spectrum (46.2%), followed by anger (23.5%), sadness (17.0%), and happiness (11.4%). By contrast, Fig. 9.6 illustrates that in far countries, anger (41.6%) and fear (27.2%) are most prevalent, with sadness (18.6%), happiness (8.2%), and disgust (4.5%) playing smaller roles.

Table 9.2 Summary of pathos categories in six languages

Category	Frequency	Percentage
Appeal to Anger	426	6.11%
Appeal to Disgust	44	0.63%
Appeal to Fear	360	5.16%
Appeal to Joy	102	1.46%
Appeal to Sadness	208	2.98%
No Pathos	5810	83.25%
Unclear/Incorrect	26	0.37%
SUM	**6976**	**100%**

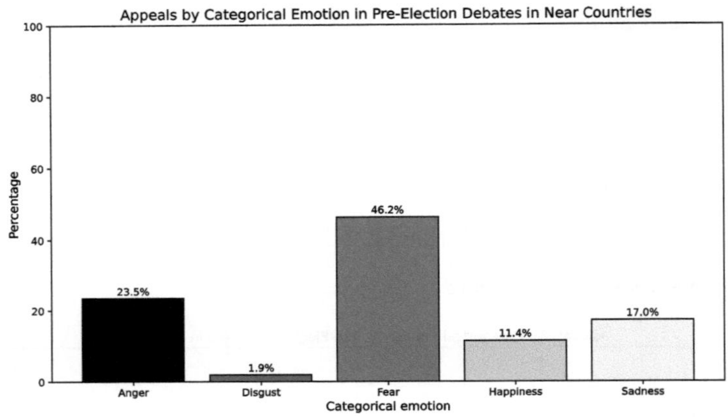

Fig. 9.5 Pathos in "Near" countries—by emotion category

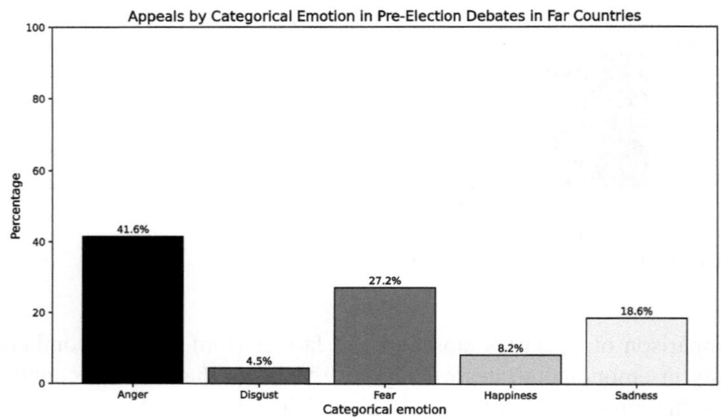

Fig. 9.6 Pathos in "Far" countries—by emotion category

9.3.2 *Case Studies in Political Discourse*

Below I present selected examples showing how appeals to emotions are executed in six different countries. This analysis shows certain trends, how pathos is universal across languages and cultures, and how Aristotle's concept from over 2000 years is still suitable for the analysis of modern discourse. At the same time, analysis reveals differences and specifics of given languages.

9.3.2.1 Anger

For anger I would use cases from three countries far from war, as those are countries in which debates anger usually dominated, as visible in Fig. 9.6. First, example (22) a case from a Greek debate, where the speaker is attempting to elicit anger towards their opponents by accusing them of acquiring wealth.

(22) Speaker: It is a shame that those who have awarded tens of millions in direct assignments and closed tenders are telling us where to find the funds to give raises to public sector employees and pensions to retirees.[26]

We can see that almost identical strategy of eliciting anger by contrasting rich and poor, big and small, can be observed in Dutch debate in example (23), where the speaker compares the levies imposed on household energy consumption with those used for the industry. Again, the aim here is to elicit anger in the audience and direct it towards those responsible for such situation—current and previous government.

(23) Speaker: But above all, households pay twelve times as much energy tax, as the large, energy-guzzling companies.[27]

Not surprisingly, we find a very similar frame of anger appeal in the text of Spanish debate. Again, the speaker paints a picture of a desired good, something that every human needs—housing, contrasted with offshore funds buying them, and of unspecific "some" profiting. To make the appeal to anger stronger, the speaker uses the phrase "vulture funds", which in Spanish, just like in English has very negative connotation to scavengers, and is used to describe funds buying debts. Regardless of speakers knowing the exact meaning, the phrase itself can elicit anger towards those responsible for such situation.

(24) Speaker: The housing law, your policy has always been to liberalize land so that some can profit and to sell public housing to vulture funds.[28]

---

[26] [Gr.] Είναι ντροπή αυτοί οι οποίοι έχουν δώσει δέκαδες εκατομμύρια απευθείας αναθέσεις και κλειστούς διαγωνισμούς να μας λένε πού θα βρείτε να δώσετε αυξήσεις στους δημοσίους υπαλλήλους και συντάξεις των ταξιούχους.

[27] [Nl.] *Maar vooral huishoudens betalen twaalf keer zoveel energiebelasting... als de grote slurpende bedrijven.*

[28] [Es.] *Ley de vivienda, su política ha sido siempre liberalizar el suelo para que algunos hagan negocio y vender las públicas a los fondos buitres.*

9.3.2.2 Disgust

Disgust, again in the context of public pre-election debate might be considered a surprising emotion, if we think about Ekman's classic, physiology-related definition presented in Part II of this book. Similarly, in terms of emotion-eliciting words presented in lexicons, we would not expect "slime" or "vomit" to appear in these debates. Therefore it is even more interesting to observe how speakers engage abstract topics of political debates into language formulations aiming at eliciting disgust—but culturally entrenched disgust.

Let us first examine an example (25), from Estonian debate, where the speaker is appealing to disgust. This disgust, however, seems to be mixed with anger in here. When we turn to Plutchik's emotion wheel, we can observe that combination of disgust and anger is treated as a dyad, a complex emotion of contempt. In this context, the sentence expresses speaker's outrage. It implies that the speaker views the current form of the distribution of wealth as not just wrong but fundamentally shameless or audacious in its inappropriateness. The emotional weight aligns with strong disapproval or even scorn and is based on right-wing values, painting current government policy as wasteful.

The Estonian word "häbematus" used in the original example consist of the root "häbe", meaning "shame" or "modesty", and the suffix "-matus" indicating a lack of something. Therefore the direct translation would be closer to the word "shamelessness". In English, in the context of the political debate, we might expect the use of the word "abomination", also revealing the mixture of disgust and anger, summarizing in contempt.

One problematic element of this example lies in its clear reliance on emotional expression, the speaker is showing their stance towards an issue. Is it also an emotional appeal? In the context of a pre-election debate, the speaker likely makes this claim not merely to vent but to serve a deliberate persuasive purpose: attacking their opponent. Such rhetorical moves are common in political argumentation, where speakers often aim to provoke emotional responses in their audience to bolster their stance or undermine opposition. In this case, as seen in example (25), the speaker combines the expression of personal emotion with an invitation for the audience to share this sentiment. The speaker frames the current government's policies as wasteful and an abomination, appealing to a shared sense of moral outrage. This strategy encourages the audience to align with their viewpoint and reject the policies or values of their opponents.

(25) Speaker: The distribution of wealth, that is an abomination.[29]

Another example, (26), coming from Polish debate, contains and interesting mixture of eliciting disgust based on moral values, performing *ad hominem* attack at other candidate, and wrapping it all in rhetorical question. First, the speaker uses a strong word—"coward" and contrasts it with the position of "prime minister".

[29] [Et.] *Jõukuse laial jagamine, see on häbematus.*

(26) Speaker: I therefore turn to you: do you want a coward as your prime minister?[30]

In example (26), just like in the previous example, the emotion of disgust to be elicited here is not based directly on physical reactions, but more on the moral values shared by the society, which values courage as characteristic of public servants and feels contempt towards cowards. This resonates with Polish self-constructed identity as a courageous, patriotic nation. Thus the speaker is performing *ad hominem* at their opponent, and a very strong one. As Walton puts it (1992), when embellished with strong emotional appeals, *ad hominem* may become not only a character attack, but in fact a character assassination, a situation in which the attacked speaker has no rhetorical move left. If someone is a coward, how can we trust anything a coward says? Interestingly, because the speaker here performs such a strong attack, they wrap it in a form of rhetorical question, combined with turning to the audience. The audience themselves is asked whether they want to have a coward as a prime minister, a question to which an answer is an obvious one. It also implicitly reinforces the attack by framing the audience as morally responsible for the answer. The speaker does not want to own the claim that their opponent is a coward, and therefore hides behind a surface level of question, at the same time, effectively introducing the claim ("my opponent is a coward") along with its emotional appeal into the discourse. This strategy relates to the concept of "poisoning the well", as silencing the opponent, which strategy recently received an interesting new approach, starting from the claim that speakers are never neutral (Holgado and Marchal 2024).

9.3.2.3 Fear

As the emotion of fear has special place both in Walton's studies (1992) and in particular in this study of pre-election debates, I will present examples from all six studied debates. In the countries near Russian aggression on Ukraine (i.e. Estonia, Finland, Poland) the category of fear appeals is dominated by the topic of war. There are some examples referring to other potential threats (i.e. climate change, finances), however the war seems to naturally be the main topic. Therefore all examples below represent this topic.

Example (27), from an Estonian pre-election debate, represents a fear appeal focused on the dangers of disinformation. The speaker identifies disinformation as a significant societal threat, describing it as a force capable of fracturing society and sowing discord among individuals. The speaker presents the danger of disinformation as a threat to societal cohesion, creating an emotional impetus for listeners to recognize its risks. Implicitly, the speaker suggests the role of Russia in active sowing of disinformation in other countries, with the aim of weakening them by creating social conflicts. As most citizens of Russia-neighboring countries are

[30] [Pl.] *Zwracam się zatem do Państwa: czy chcecie tchórza za premiera?*

aware of these practices due to the longstanding public information campaigns, the speaker taps into this cultural context to elicit fear.

(27) It reawakens that interest again, and what must be done constantly is to talk about the danger of disinformation, why it is dangerous, because it breaks apart our society, it drives our people into conflict with each other.[31]

Example (28), from a Finnish pre-election debate, also illustrates the use of fear appeals, focusing on the hostility of Russia as a dangerous actor. The speaker frames Russia as actively seeking to cause harm and generate trouble, positioning the situation as a burden of war that Finland must endure.

(28) Speaker: In my opinion, it would be more worth thinking about the issue this way: since Russia is clearly a hostile actor and surely has the desire to create various kinds of trouble and cause harm, we are soldiering the burden of the war.[32]

The speaker presents the key components of fear appeal: a clear and immediate threat (Russia's hostility) and the implicit need for action or vigilance in response. The emphasis on Russia's behavior serves to heighten the audience's sense of urgency and the necessity for unity and preparedness, subtly guiding them toward the speaker's perspective.

Example (29) from Polish pre-election debate, represents a case of fear appeal, which wile not ideally matching to the scheme introduced by Walton et al. (2008), contains main elements of this scheme. First, the speaker is presenting a potential threat (Russian soldiers crossing Polish border). This is a very strong image, something at the time of near Russian aggression would be a strong impuls for eliciting fear in hearers. But the speaker is contrasting it with the guarantee of safety—the election of new democratic government, and in particular The Third Way—a party represented by the speaker. Here we have then a typical structure of fear appeal scheme—the speaker presents themselves as the only way to prevent a dangerous event from happening.

Example (29) also includes purposeful choice of words. In the English translation, the term "żołdak" is rendered as "soldier" for clarity and simplicity, as there is no direct equivalent in English that captures the historical and pejorative details of the Polish word. However, "żołdak" in Polish carries a strong negative connotation, historically referring to undisciplined, ruthless, or mercenary soldiers. Its use in this context reflects the speaker's intention to elicit fear by emphasizing the brutality and aggression associated with such figures, rather than using the neutral term "żołnierz" ("soldier").

(29) Speaker: The Third Way and the future democratic government are a guarantee that not a single Putin's soldier will step even a meter beyond the Polish border.[33]

---

[31] [Et.] *Äratab jälle seda huvi, ja mida peab kogu aeg tegema, rääkima selle desinfo ohu mõtet, miks see on ohtlik, sest see lõhub meie ühiskonda, see ajab meie inimesed oma vahel tülli.*

[32] [Fi.] *Minun mielestäni asiaa kannattaisi ajatella pikemminkin niin päin, että kun Venäjä on selkeästi vihamielinen toimija ja sillä varmasti on halu tehdä erilaista kiusaa ja aiheuttaa.*

[33] [Pl.] *Trzecia Droga i przyszły, demokratyczny rząd to gwarancja, że żaden żołdak Putina nie wejdzie ani na metr za polską granicę.*

Appeals to fear were also present in the debates in countries further from Russian aggression: Greece, the Netherlands, and Spain, but they were in general less frequent and less uniform.

For Greece I would like to present two examples of appeals to fear, as they show two distinct strategies, both interesting for our topic. Example (30) shows how speakers are exploring the emotional power of metaphor, the topic I have explored already in Part II. The speaker is talking about the problem of economic mismanagement, however they are using the vocabulary related to the violent images of a battlefield—"walking into a minefield" and "without defenses", to increase the emotional power of the warning.

(30) Speaker: Our country is walking into a minefield of economic mismanagement, without defenses against what is coming.[34]

In example (31), the situation reflects a deeply rooted and complex issue, the Greek-Turkish tensions over disputed territories. In this case, the speaker does not rely on metaphor or abstraction; the direct and vivid depiction of a potential military incident is sufficient to elicit fear. For an audience familiar with the historical and geopolitical context, the possibility of Turkey's reaction and the appearance of frigates evokes a palpable sense of urgency and insecurity.

(31) Speaker: There is a strong possibility that Turkey will react, and two frigates will appear there.[35]

These two examples illustrate how appeals to fear in Greek debates can use distinct strategies—one through metaphorical imagery to amplify economic concerns, and the other through direct reference to longstanding geopolitical tensions. This shows how fear appeals are not uniform in their nature and can be used in various forms in addressing diverse national anxieties.

While Dutch debate contains references to geopolitical situation and military tensions, such as Israel and Hamas, certain appeals to fear include what has been present also in lexicons of emotive words presented in Part II of this book—references to illness, as in example (32).

(32) Speaker: People living in the area just get cancer from it.[36]

For Spain, the fear appeals identified often included socio-political instability, crime, economic inequality, and threats to democracy, resting on the audience's concerns and anxieties, just like in example (33). The speaker uses the Spanish word "grave" (meaning "serious") to elicit fear, to add to the importance of the fact they mention.

---

[34] [Gr.] Η χώρα μας βαδίζει σε ένα ναρκοπέδιο της διευθύνσης οικονομίας, χωρίς άμυνες απέναντι σε αυτό που έρχεται.

[35] [Gr.] Είναι ισχυρό το ενδεχόμενο η Τουρκία να αντιδράσει και να εμφανιστούν δύο φρεγάτες εκεί.

[36] [Nl.] *Mensen die daar in de omgeving wonen, die krijgen daar gewoon kanker van.*

(33) Speaker: Today in Spain, more people die from heatwaves than in traffic accidents, and that is a very serious fact.[37]

9.3.2.4 Joy

Appeals to joy (or happiness) are not very frequent in our data, and they mainly refer to the future, they fit roughly into the argument from positive consequences (Walton et al. 2008)—"vote for me, and the good things will come". Again, in this case, we can observe several patterns repeating across cultures, but also certain specific differences. Example (34) from Netherlands illustrates a forward-looking appeal to joy, where the speaker envisions a future in which citizens collectively acknowledge the achievements of the country. The phrase "what we're doing well" evokes a sense of pride and contentment, located in the future.

(34) Speaker: So that everybody can say, later on, what we're doing well in this country.[38]

There are also other strategies, less straightforward, as visible in example (35). In this Finnish example, the appeal to joy is interwoven with expressions of hope in a difficult context. The speaker frames the desired outcome, successful negotiations and the release of hostages, as a source of relief and happiness. While the immediate situation is fraught with tension, the reference to hope provides an emotional anchor, encouraging the audience to remain optimistic and united in their support for the speaker's efforts or policies. This blend of hope and joy is particularly impactful, as it connects with the audience's empathetic response to a high-stakes scenario.

(35) Speaker: I hope that the negotiations are successful and the hostages are released.[39]

This Spanish example (36) employs a more abstract appeal to joy by calling the act of voting a "vote of hope". The speaker suggests that such a vote is motivated by optimism for a better future. By linking unity with hope and joy, the speaker not only appeals to positive emotions but also positions their message as the pathway to fulfilling these aspirations.

(36) Speaker: But if any vote is the vote of hope, this is the vote for unity.[40]

---

[37] [Es.] *A día de hoy en España mueren más personas por olas de calor que no en accidentes de tránsito y eso es un dato muy grave.*

[38] [Nl.] *Zodat ook iedereen straks kan zeggen, wat hebben we het goed voor elkaar in dit land.*

[39] [Fi.] *Toivon, että neuvottelut ovat tuloksekkaat ja panttivangit pääsevät vapaaksi.*

[40] [Es.] *Pero si algún voto es el de la esperanza, este es el de sumar.*

9.3.2.5 Sadness

One might wonder, how appeals to sadness could be useful for the rhetorical aims of speakers in pre-election debates. Does eliciting the feeling of loss fit into the arguing for election? As we can see, speakers in pre-election debates can use non-direct strategies to use sadness appeals to their aims.

An example, perfectly fitting into the loss of valued object frame, comes from Finnish debate in which speaker is bringing up the passing of Martti Ahtisaari, a Peace Nobel Price laureate, and a person widely cherished in the society. The speaker is using this reference to slow down the heated debate about military actions, both in terms of Israeli-Hamas conflict as border security of Finland.

(37) Speaker: Finally, we also remember the founder of CMI, Martti Ahtisaari, who was buried on Sunday.[41]

A Spanish example (38) makes a stark and emotionally charged appeal to sadness by presenting the grim reality of suicide rates—11 suicides per day in the country. This statistic is not merely a fact but a deliberate invocation of loss and despair, aimed at stirring the audience's empathy and moral responsibility. The speaker is using sadness to stress the urgency of addressing mental health issues.

(38) Speaker: There are 11 suicides per day in this country.[42]

In example (39) the speaker is attempting at eliciting feeling of abandonment in the audience. The phrase "we have felt alone in this fight" conveys a sense of isolation and vulnerability, inviting the audience to share this with speaker and symphatize with them.

(39) Speaker: And many times, we have felt alone in this fight, I have to say it clearly.[43]

9.3.2.6 Conclusions and Limitations for Multi-language Debates Study

The focus of this multi-language debates study were fear appeals observed in language. While we haven't tested fear appeals experimentally ourselves in my team, next to the argumentation studies literature on fear appeals (mainly Walton 1995), there is also a large body of experimental work on appealing to fear, starting with the classic study by Janis (1967). Janis proposed the Gauss distribution-based Fear Drive Model to explain varying human responses to fear. According to this model, moderate fear can motivate individuals to address risks, whereas excessively low or high fear levels might inhibit such actions. More recent research points to the complexity in these relations: researchers (LaTour and Tanner 2003; Ali

[41] [Fi.] *Lopuksi muistamme myös sunnuntaina haudattua CMIn perustajaa Martti Ahtisaarta.*

[42] [Es.] *Hay 11 suicidios al día en este país.*

[43] [Es.] *Y muchas veces nos hemos sentido solos en esta defensa, lo tengo que decir así de claro.*

et al. 2019) observed that positive aspects of fear can enhance preventive measures. They demonstrated this dynamic in contexts such as radon gas contamination, showing a strong link between fear and preventive actions. Another study (Hartmann et al. 2014) found that heightened fear of environmental threats tends to increase engagement in pro-environmental behavior. However, the relationship between fear and preventive actions remains highly context-dependent. This shows that while our study results point to the differences in the frequency of fear appeals, it is still difficult to predict a uniform effect of those appeals on audience.

One limitation of this study lies in its cultural scope. While I analyzed six European languages, which reveal considerable differences between cultures, such as those of Estonia, the Netherlands, and Spain, the study does not extend to non-Western cultures or languages. The methodologies proposed here are intentionally straightforward, designed to investigate emotional appeals in language. Interestingly, the findings suggest that some insights dating back to Aristotle remain relevant today, indicating universal patterns across languages that can often be interpreted using frameworks such as Plutchik's wheel of emotions. At the same time, cultural differences are significant and require perspective of researchers representing relevant cultural circle. I believe it is essential for researchers from various cultural and linguistic backgrounds to take an active role in conducting such analyses. Although large language models provide unprecedented opportunities to study languages we may not personally speak, the human factor, particularly an understanding of cultural context, remains crucial in analyzing emotional appeals effectively. Future research applying this methodology to languages from non-Western cultural circle would not only deepen our understanding but also validate its broader applicability across diverse cultural frameworks. I hope this work inspires further studies in this area.

Another limitation comes from the use of Large Language Model—GPT-4. Like all large models, GPT-4 is non-deterministic and prone to generating erroneous outputs. Nevertheless, I believe that the methodological rigor and human oversight employed in this study mitigate these issues to a significant extent. While pathos mining (i.e., the identification of emotional appeals) and the translation of data from various languages were performed automatically, all results underwent human inspection. Although the method, as mentioned, is far from perfect, the agreement between the model's predictions and human evaluations is comparable to the inter-annotator agreement typically observed between different human coders. GPT-4, like other LLMs, is not a general artificial intelligence but a system highly specialized in processing and generating language. While not infallible, its predictions and outputs, when understood within this framework, can be trusted for linguistic analysis when paired with proper methodological safeguards.

In terms of chapter summary, the studies have demonstrated that Large Language Models (LLMs) can be effectively used to identify emotional appeals in natural language argumentation across diverse contexts and languages. In two complementary studies, competitive student debates and multilingual political debates, I applied LLMs to annotate pathos appeals, evaluating distribution of emotion categories. In competitive debates, students, without being prompted or trained in rhetorical theory, used a wide range of emotional appeals, confirming the intuitive and pervasive nature of pathos in persuasive speech. In political debates, we observed that proximity to military conflict significantly shaped the frequency and type of emotional appeals, particularly fear. These results align with predictions from argumentation theory and psychology of emotions and show that LLMs, while imperfect, can detect rhetorical patterns consistent with both empirical and theoretical expectations.

At the same time, the chapter has listed important methodological limitations and conceptual challenges. First, while LLMs proved to be surprisingly robust in replicating human annotation patterns, their context sensitivity remains limited, especially in detecting subtle rhetorical strategies or interpreting emotional sequences that unfold over several utterances. This is a limitation that they share with older methods, such as lexicons. In addition, certain emotion categories, such as disgust or mixed emotional appeals, pose difficulties for both human and machine annotators. These challenges point to the need for hybrid methodologies that combine computational scalability with qualitative insight and cultural competence. Moreover, we are reminded that LLMs do not feel emotions or understand rhetorical intentions as humans do; they respond to patterns in linguistic input. However, their utility lies in supporting scalable, replicable and theoretically grounded analysis of pathos, especially when integrated with careful prompt design and human oversight. With this final application of LLMs to pathos mining, this book completes its interdisciplinary journey. From Aristotle's original conception of pathos as one of the means of rhetorical persuasion, through psychological models of emotion and their experimental operationalizations, to the computational implementation of pathos recognition at scale, I have hopefully shown how emotional appeals can be systematically studied in natural language argumentation. The Model of Interactional Pathos in Argumentation (MIPA), developed and applied throughout the book, offers a conceptual and methodological bridge linking rhetorical theory, cognitive science, and artificial intelligence. It demonstrates that pathos is not a vague or intangible rhetorical flourish, but an observable, analyzable phenomenon grounded in the interplay between speaker intention and audience reaction.

References

Ali K, Zain-ul abdin K, Li C, Johns L, Ali AA, Carcioppolo N (2019) Viruses going viral: Impact of fear-arousing sensationalist social media messages on user engagement. Science Communication 41(3):314–338, https://doi.org/10.1177/1075547019846124

Gajewska E, Budzynska K, Konat B, Koszowy M, Kiljan K, Uberna M, Zhang H (2024) Ethos and pathos in online group discussions: Corpora for polarisation issues in social media. arXiv preprint arXiv:240404889 2404.04889

Hartmann P, Apaolaza V, D'souza C, Barrutia JM, Echebarria C (2014) Environmental threat appeals in green advertising: The role of fear arousal and coping efficacy. International Journal of Advertising 33(4):741–765, https://doi.org/10.2501/IJA-33-4-741-765

Holgado MA, Marchal AH (2024) Against the neutral view of poisoning the well. Argumentation https://doi.org/10.1007/s10503-024-09646-9, https://link.springer.com/article/10.1007/s10503-024-09646-9

Janis IL (1967) Effects of fear arousal on attitude change: Recent developments in theory and experimental research. Advances in Experimental Social Psychology 3:166–224

Kiljan, K., & Konat, B. (in press). War proximity as a catalyst for variations in emotional appeals distribution in pre-election debates. Central European Journal of Communication.

LaTour MS, Tanner JF (2003) Radon: Appealing to our fears. Psychology and Marketing 20(5):377–394

Obrębska, M and Konat, B and Kleka, P and Kiljan, K and Dembska, N (in press) Personality traits and argumentation style as factors in perceived argument strength in natural conditions of competitive debating. The role of psychological matching of speaker and listener. Available at SSRN: https://ssrn.com/abstract=5416998 or http://dx.doi.org/10.2139/ssrn.5416998

Plutchik R (2001) The nature of emotions: Human emotions have deep evolutionary roots, a fact that may explain their complexity and provide tools for clinical practice. American scientist 89(4):344–350

Plutchik R (2003) Emotions and life: Perspectives from psychology, biology, and evolution. American Psychological Association

Walton D (1992) The Place of Emotion in Argument. The Pennsylvania State University Press, University Park, PA

Walton D (1995) Appeal to pity: A case study of the argumentum ad misericordiam. Argumentation 9(5):769–784

Walton D (2013) Scare tactics: Arguments that appeal to fear and threats, vol 3. Springer Science & Business Media

Walton D, Reed C, Macagno F (2008) Argumentation Schemes. Cambridge University Press, Cambridge, https://doi.org/10.1017/CBO9780511802034

Conclusion

This book set out to answer a foundational question: how can emotional appeals in natural language argumentation be systematically identified, modeled, and interpreted using insights from rhetorical theory, psychology, and computational linguistics? The response was developed across three integrated parts, each building upon the last. The result is a novel framework for understanding and analyzing pathos, not as a secondary or ornamental feature of persuasion, but as an interactive, empirically measurable component of argumentative discourse. The Model of Interactional Pathos in Argumentation (MIPA), developed throughout the chapters, provides a conceptual and operational structure that connects speaker intention and audience response through analysis of emotional reactions.

The book pursued this question through a multimethod approach. In Part I, I examined rhetorical and pragmatic theories of pathos, then applied them in manual corpus studies that annotated emotional appeals using argument schemes. In Part II, I drew on psychological theories of emotion to explore experimental approaches and emotion lexicons, establishing empirical ground for interpreting how emotional language functions across individuals and populations. Finally, Part III presented computational applications: first, lexicon-based pathos mining and sentiment analysis of audience reactions; then, large language model (LLM)-based methods for automatic pathos detection. These three parts, and their respective chapters, together demonstrate that emotion in argumentation can be rigorously studied using interdisciplinary, empirical tools.

Several key insights arise. Emotional appeals are frequent and diverse, appearing even in spontaneous argumentation by untrained speakers, such as students or social media users. Audience reactions, whether measured experimentally or in real-world settings, do not always mirror the speaker's intended emotion. Instead, responses may intensify, mitigate, or shift the emotional valence in unexpected ways. Political context also matters: appeals to fear are more prevalent in debates closer to armed conflict, while anger appears consistently across national settings. Particularly in social media discourse, positive emotional appeals, though less frequent, correlate

B. Konat, *Emotional Appeals in Argumentation*, Argumentation Library 47,
https://doi.org/10.1007/978-3-032-06332-8

with reduced antagonism and greater engagement, suggesting their value as a rhetorical counter-strategy in polarized environments.

Underlying all of those studies, is The Model of Interactional Pathos in Argumentation (MIPA), which defines pathos as an emotional appeal performed in a persuasive interaction by the speaker with the aim of eliciting emotions in the audience in order to increase the perceived strength of the speaker's argumentation. It draws on rhetorical theory, speech act theory, and psychological models of emotion to clearly distinguish between the speaker's intention to provoke emotion and the audience's emotional response. By conceptualizing this as a dialogical, cognitive, and linguistic process, MIPA allows for the empirical study of emotional appeals across both experimental and computational approaches. The model has been used to annotate pathos in corpora, to test elicitation and reaction in controlled settings, and to develop computational detection methods. This structure provides a consistent analytical approach across disciplines and helps clarify distinctions often overlooked, such as that between emotional expression and emotional elicitation.

The broader contribution of this work lies in positioning pathos as a research object equally suited for qualitative analysis and computational modeling. MIPA operationalizes the classical Aristotelian notion of pathos in a contemporary, interdisciplinary context. It allows for tracking emotional influence across dialogical turns, for analyzing appeals in multilingual settings, and for quantifying their effects on audience response. As such, this book contributes to a growing body of work that sees emotion not as a threat to rationality, but as a constituent feature of communication, especially in persuasive, deliberative, and political discourse.

While the studies presented confirm the utility of computational methods for pathos analysis, they also reveal clear methodological boundaries. Lexicon-based approaches lack contextual sensitivity and often overgeneralize emotion categories. Sentiment analysis models, though effective in tracking large-scale emotional expressions, struggle with ambiguous or ironic language. LLM-based annotations show promise in aligning with human judgments, but remain dependent on prompt design and domain adaptation.

Across all methods, a recurring challenge is the distinction between emotion-eliciting language and expressed emotion, a distinction that is theoretically crucial but computationally demanding. With the introduction of MIPA and its application to both experimental and computational approaches, I hope to help clarify this distinction and reduce the persistent confusion between emotional appeals and emotional reactions. Recognizing and modeling this difference is needed for advancing both theoretical insight and practical analysis.

Looking ahead, several directions for future research can be identified. First, advances in LLMs offer new possibilities for modeling dynamic emotional exchanges, especially in dialogical contexts where roles of speaker and audience shift rapidly. Second, integrating multimodal data, such as facial expressions, tone, and physiological signals, will improve the ecological validity of computational models. However, this direction has been deliberately excluded from the present research as it would have extended the scope too far beyond the focus of the book. Multimodal emotion recognition constitutes a distinct subfield with its own theoretical and methodological foundations, which deserve dedicated treatment elsewhere. Third, there is an urgent need to explore the ethical dimensions of automatic pathos detection, particularly in areas such as political communication, journalism, and online moderation. Finally, continued collaboration across fields: argumentation theory, cognitive science, NLP, and ethics will be essential to responsibly advance the study of emotion in public discourse.

This book has aimed to trace a trajectory: from Aristotle's foundational conception of pathos, through psychological experimentation, to scalable computational models rooted in both theory and data. By doing so, it hopes to contribute to a deeper understanding of emotional dynamics in argumentation and to provide tools for scholars, educators, and practitioners navigating a world in which persuasion is inherently emotional and interactive, and where each of us alternates between the roles of rhetor and audience.

Index

A
Argumentation, 7
  definitions of, 7
  fallacious, 16
  in Inference Anchoring Theory, 38
  mining, 97–103
  in Pragma-Dialectics, 35
  and pragmatics, 33
Aristotle, 3–6, 41

D
Debate
  competitive, 88, 129
  political, 45, 110, 142
  in social media, 116

E
Emotional appeals, *see* Pathos
Emotions
  basic, 67
  categorical, 67
  Ekman's model of, 66
  naming of, 53–55
  Plutchick's model of, 67

G
Gilbert, M.A., 24–27

H
Hundleby, C., 23

L
Lexicons of emotions
  computational applications, 110
  psychological foundations, 73

M
Macagno, F.
  emotive words, 19
Manipulation
  definition of, 10
Metaphor, 57–59

P
Pathos
  in argumentation theory, 16–22
  in classical rhetoric, 4–6
  in feminist and queer theory, 22–27
  interactional model of, 40, 47, 83, 88
  in manipulation, 10
  mining, 109–122
  mining with AI, 142–153
  in persuasion, 8
  in pragmatics, 32–35
  in psychological experiment, 83–90

B. Konat, *Emotional Appeals in Argumentation*, Argumentation Library 47,
https://doi.org/10.1007/978-3-032-06332-8

Persuasion
definition of, 8
in psychological research, 79
three means of, 4
Plantin, C., 54

S
Speech acts, 34–44

W
Walton, D.
argument schemes, 45
"The Place of Emotion in Argument", 23

MIX
Papier aus verantwortungsvollen Quellen
Paper from responsible sources
FSC® C105338

If you have any concerns about our products,
you can contact us on
ProductSafety@springernature.com

In case Publisher is established outside the EU,
the EU authorized representative is:
Springer Nature Customer Service Center GmbH
Europaplatz 3, 69115 Heidelberg, Germany

Printed by Libri Plureos GmbH
in Hamburg, Germany